Society of Earth Scientists Series

Series Editor

Satish C. Tripathi, Lucknow, India

The Society of Earth Scientists Series aims to publish selected conference proceedings, monographs, edited topical books/text books by leading scientists and experts in the field of geophysics, geology, atmospheric and environmental science, meteorology and oceanography as Special Publications of The Society of Earth Scientists. The objective is to highlight recent multidisciplinary scientific research and to strengthen the scientific literature related to Earth Sciences. Quality scientific contributions from all across the Globe are invited for publication under this series. Series Editor: Dr. Satish C. Tripathi

Satish C. Tripathi • Naresh Chandra Pant
Sameeta Rajora
Editors

Geoconservation and Geotourism Potential of India

The Society of Earth Scientists
2008

Springer

Editors
Satish C. Tripathi
The Society of Earth Scientists
Lucknow, Uttar Pradesh, India

Sameeta Rajora
Madhya Pradesh Ecotourism
Development Board
Bhopal, Madhya Pradesh, India

Naresh Chandra Pant
National Centre for Polar and Ocean
Research
Goa, Goa, India

ISSN 2194-9204 ISSN 2194-9212 (electronic)
Society of Earth Scientists Series
ISBN 978-3-031-81016-9 ISBN 978-3-031-81017-6 (eBook)
https://doi.org/10.1007/978-3-031-81017-6

This Springer imprint is published by the registered company Springer Nature Switzerland AG
The registered company address is: Gewerbestrasse 11, 6330 Cham, Switzerland

Series Editor's Preface

Geological heritage is like a history book which records the processes of evolution of Earth. As such keeping the record book in good condition for future generations is the responsibility of every citizen considering the fact that these records are not renewable and once destroyed cannot be recreated. The geoheritage conservation and development of geoparks has gained momentum in a large number of countries including smaller ones. As such there are 213 UNESCO Global Geoparks in 48 countries but not in India. The concept of geoheritage conservation was there in India as early as 1951 when Tiruvakkarai National Fossil Wood Park was established by the Geological Survey of India and declared a National Geological Monument. Later, a large number of geoheritage sites were added to this list mainly in the 1970s. However, geoheritage conservation took a back seat afterwards. Further, fast development is the real threat to geological heritage and most of the policymakers are unaware of any knowledge about the significance of such sites. On the other hand, geoscientific community has also failed to apprise them and make them aware. As such geoheritage conservation, despite all our efforts, is still not gaining momentum.

This book is basically on lesser known geoheritage sites and potential geoparks of India written in semi-scientific language. This is in addition to all our efforts to bring geoheritage conservation on the central stage. We realise our duty and continue to strive to reach out to all stakeholders to meet the requirements so that India sees geoconservation and geoparks in the forefront. I am thankful to the editors and contributors of this volume for extending support towards these efforts.

Lucknow, India

Satish C. Tripathi

Contents

About the Editors

Satish C. Tripathi is former Dy. Director General, Geological Survey of India; founder general secretary of the Society of Earth Scientists; Editor, e-Journal *Earth Science India*; and Series Editor, Society of Earth Scientists Series by Springer. Dr. Tripathi contributed immensely to the geology of Son-Narmada Lineament Zone, including detailed study of dinosaur fossil bearing sedimentary succession of Madhya Pradesh. His contributions on the geology of Lesser Himalaya and Trans-Himalaya of Spiti-Ladakh are widely referred. He has published over 34 research papers in national and international journals and edited two books. He has organised over 12 national and international conferences and field workshops. Of late Dr. Tripathi started working on geological heritage conservation and development of geoparks in India. In 2019, he organised 'Walk to Save our Geoheritage' in six different cities across India, and submitted Status Report and Draft Legislation for the conservation of geoheritage to the Government of India. He is credited with bringing Indian geoheritage sites to the international stage, and with his efforts, the IUGS has recognised Indian geoheritage sites under First 100 and Second 100 IUGS Geoheritage Sites. Dr. Tripathi is currently involved in the establishment of geoparks in Bagh, Chitrakoot, Zawar, and Ramgarh with the support of State governments.

Naresh Chandra Pant is former Senior Professor of geology in the University of Delhi. He has a total academic and research experience of over 41 years in which over 26 years was with professional institutions (ONGC and Geological Survey of India). Prof. Pant has published over 110 research papers and edited four books with significant research outputs in the form of establishing supercontinent reconstruction, described high pressure (HP) metamorphism for the first time from an Indian Archean craton (Bundelkhand) implying processes similar to present-day subduction, and applied innovative and uncommon application of EPMA for estimating the bulk composition of selected micro-domains which is essential for understanding phase transformations where chemical systems are of few millimetres scale, inferring sub-ice geology of east Antarctic shield from proximal marine sediments and many others. He has made major contribution in establishing, developing, and nurturing micro-domain characterisation laboratories in India and propagating the chemical geochronology in the country. He has received prestigious National Mineral Award and Antarctic Silver Jubilee Award. He is a Fellow of Electron Microscope Society of India. Prof. Pant is involved in the management and development of the Indian Antarctic Programme for over a decade, besides being one of the Chief Officers of the Geosciences Group of the Scientific Committee on Antarctic Research (SCAR) for the similar duration.

Sameeta Rajora is an officer of the 1992 batch of the Indian Forest Service borne on the Madhya Pradesh Cadre. She is the Chief Executive Officer of Madhya Pradesh Ecotourism Development Board and an expert on community-based ecotourism and conservation outreach programmes. She has a Ph.D. in Botany and her extensive work is on sacred groves protected by the Bhil Community. Dr. Rajora has a vast experience of over 30 years in the field of forest and wildlife management. She has successfully designed and anchored the largest conservation education programme 'Anubhuti' with an outreach to about 3 lakh participants, mostly school children. Dr. Rajora's interest towards fossil conservation began in the year 1999 when she was Divisional

Forest Officer, Dhar, where she first saw the Bagh dinosaur, wood, and marine fossils. Ever since she has been creating awareness and working for the protection of fossils. The Madhya Pradesh Government declared the fossil-rich forest area of 89.4 hectare with dinosaur nesting sites as ‘Bagh Dinosaur National Park’ in the year 2011. She has co-authored a paper on the study of Dinosaur Fossil National Park, Bagh region, to develop it as a UNESCO global geopark published in 2023 in the *International Journal of Geoheritage*.

Geoconservation and Geotourism Potential of India

Satish C. Tripathi, Naresh Chandra Pant, and Sameeta Rajora

Abstract India is studded with spectacular geoheritage sites and have every potential to be developed as geoparks. However, geoconservation and geoparks are still lesser known terms particularly for policy makers and leaders. There a large number of geoheritage sites of global significance are in dilapidated condition and there is no policy and legislation to conserve these sites. The result is that there is no Geopark in India where as it has 42 World Heritage Sites (but not on geoheritage basis). In this paper we have provided current status of geoheritage conservation in India and future scope to develop geoparks along with summary of the papers presented in this volume.

Keywords Geoconservation · Geopark · Geoheritage sites · Geoheritage legislation

1 Introduction

'*Heritage is our legacy from the past, what we live with today, and what we pass on to future generations. Our cultural and natural heritage are both irreplaceable sources of life and inspiration*' (UNESCO). Heritage has both living as well as non-living components and in the human context, it is often distinguished as cultural and natural. There are problems with this also as there are overlaps between the two as seen in the UNESCO definition which incorporates these two attributes in an overlapping manner (Azzopardi et al. 2022). These authors (Azzopardi et al. op cit.)

S. C. Tripathi (✉)
The Society of Earth Scientists, Lucknow, India

N. C. Pant
National Centre for Polar and Ocean Research, Goa, India

S. Rajora
MP Ecotourism Board, Bhopal, India

S. C. Tripathi et al. (eds.), *Geoconservation and Geotourism Potential of India*, Society of Earth Scientists Series, https://doi.org/10.1007/978-3-031-81017-6_1

have developed a new framework, which incorporates multiple dimensions of heritage values of the natural environments including their Life Framework to overcome the issues of overlapping valuation. Geoheritage refers to the geological diversity of any area, which have geological value worthy of preservation for future generations. Ingrained with the geoheritage is the idea of conservation as preservation for future is an intrinsic part of any heritage. Often on account of spectacular visual attraction of the geological heritage or of the rare science implications (*e.g.* ages in terms of millions or billions of years, extinct fossils, unique minerals etc.) these geological heritage are potential sites of tourism. Thus, the conservation and tourism are intrinsically related. Even in the practically uninhabited and remote landscapes of Antarctica, geoheritage and its conservation forms an important aspect within the Antarctic Treaty System with a system and provision for declaring Antarctic Specially Protected Areas (ASPA) and Antarctic Specially Managed Areas (ASMA) (details at https://www.ats.aq/e/protected.html) and Scientific Committee on Antarctic Research or SCAR with a dedicated Expert Group on Geoconservation (https://scar.org/science/geo/geoconservation). Even in this icy wilderness, there has been an upsurge in tourism mainly on account of its geotouristic elements since the year 2000.

The Indian subcontinent is unique in having a variety of geoheritage sites which span from some of the oldest on planet Earth to the youngest and those manifest at sea-level to some of the highest elevations on the Earth. Several of these are not only bestowed with significant scientific importance but are also spectacular in appearance. For various reasons, there have been a paucity of attention not only on the conservation of several of these geoheritage sites but their geotouristic potential has also been largely untapped. The 2500 km long Himalayan mountain chain not only possesses the rugged beauty but is also the only major young and evolving result of converging continental plates on Earth. The separation and thousands of kilometer flight of the subcontinental landmass from southern hemisphere located supercontinent Gondwanaland is well-archived in varied geological formations including large lava outpourings (Deccan Traps), incursion and retreat of seas, formation of some of the largest coal deposits, proliferation and demise of dinasaurs and other flaura and fauna, evidence of past glaciations and many other. The still continuing northward movement of the Indian plate (Molnar and Tapponnier 1977; Jade et al. 2017) has not only left its imprints in the geological records but also has impacted climate change and the evolution of Indian Summer Monsoon (ISM). The present volume is a small effort in presenting some of these geoheritage sites with emphasis on their popular science related significance and thereby signaling the possibility of these sites to be of geotouristic significance.

In the following paragraphs, a brief introduction to the some of the lesser known geoheritage sites is provided. We hope that this small beginning will lead to enhanced conservation of these sites, will help the local populations and civil governments in harnessing their geotourism potential and recognition of numerous such locales for sustainable development.

2 Geodiversity of India

Geodiversity has been defined as 'the natural range (diversity) of geological (rocks, minerals, fossils), geomorphological (landforms, topography, physical processes), soil and hydrological features. It includes their assemblages, structures, systems, and contributions to landscapes' (Gray 2013). As such, whatever abiotic we see in our surroundings is nothing but the Geodiversity. Biodiversity thrives on geodiversity and geoconservation benefits biodiversity, however, geodiversity is yet to get proper attention and recognition. Gray 2018 described 25 major 'geosystem services' coming out of our geodiversity. Although these services are for the well-being of mankind, they are yet to bring on popular domain and as such society utilises them to a great extent but do not recognise their significance. Therefore, this concept and its importance should be brought out to the public. The declaration of International Geodiversity Day by UNESCO is a good attempt to make its way into policymaking and general attention to society about its dependence on natural ecosystems. The geological and resultant geomorphological structures are important features of our geodiversity. We briefly describe them in the Indian context.

2.1 *Geomorphological Set up*

Geomorphology is the surface manifestation of the underlying geological setup. Landforms provide information on the surface processes that led to their formation and are also indicative of the climate persistent during the geological past. The land surface is a platform where all human activities take place. Interestingly, some of the landforms also have spiritual, cultural, or historical value.

The 'V' shaped southern Indian peninsula is bounded by oceans and the tectonically active young Himalayan mountain belt separates it from Asia on the north. As such, it attained the status of an independent geographic entity- the Indian subcontinent. India contains almost all geomorphic features present on the surface of the Earth like the higher Himalayan belt, the plateau of the Deccan Traps and Vindhyans, a vast alluvial tract of the Ganga river system, and long and narrow coastline, a vast tract of deserts, mangroves, lagoon, and active volcanoes. Geomorphologically, India can broadly be classified into the following groups:

1. Precambrian Terrane of peninsular India
2. Himalayan mountain belt and Indus Suture Zone
3. Indo-Gangatic Alluvium
4. Hot and cold deserts of western India and Ladakh
5. Plateau of Deccan Trap and Vindhyan sedimentary basin
6. Coastal geomorphological region

A large part of southern and western India is occupied by deeply eroded Precambrian granitoids and metamorphic sequences. Various balancing rocks developed on

a. b.

Fig. 1 (**a**) Meter-scale columnar joints in Deccan Traps of Bharudpura, Dhar, M.P.; (**b**) Fanning columnar joints of Deccan Traps, Singhana, Dhar, M.P

granitoid terrane are the center of attraction. But, largely it is monotonous on account of landforms developed. The active mountain chain of the Himalaya is marked by fast erosion coupled with continuous upliftment due to continent-continent collision. South of the Himalaya, there is a vast and fertile alluvial plain of the Ganga river system originating from the Himalaya. Fast erosion of the Himalayan rock sequence gave birth to thick alluvial deposits in the foredeep. The almost undisturbed Deccan Volcanic Province in central and western India and the Vindhyan sedimentary basin bordering Bundelkhand granitoids towards the south and the west form a large 'Plateau' and associated landforms. Columnar Joints of the Deccan Traps provide a unique landscape (Fig. 1a, b) and are places of interest. India has a very long coastline with the Bay of Bengal to the east and the Arabian Sea to the west. Whereas the Western Ghat is marked by a high mountain range, the eastern coast is the region of low-lying hills. Coastal landforms have a significant geotourism potential.

2.2 Geological Set up

The geology of India is diverse, but a large part is constituted by Precambrian igneous and metamorphic rock, which are also well known for their economic potential. Valdiya (2016) classified the Indian subcontinent into as many as nine geological terranes, each having its own lithological-petrochemical characteristics, structural layout, and evolutionary history.

1. Indian Shield with Archaean Cratons
2. Early Proterozoic Mobile Belts
3. Continent-Interior Later Proterozoic Sedimentary Basins
4. Gondwana Basins
5. Cretaceous Volcanic Provinces
6. Pericratonic Sedimentary Basins of Peninsular India

7. Himalaya Mobile Belt
8. Indo-Gangetic Plains and Quaternary Cover of Peninsular India
9. Andaman Island Arc and Central Myanmar Plain

The peninsular Indian shield is constituted by four large cratonic blocks: the Dharwar Craton in southern India; the Bastar Craton in central India; the Singhbhum Craton in eastern India and the Bundelkhand Craton in northern India. The cratonic blocks have a Palaeoarchaean nuclei (3310–3560 Ma) comprising granitic gneisses and mafic-ultramafic complexes. The cratons are separated from each other by Early Proterozoic mobile belts: Eastern Ghat Mobile Belt, Satpura-Aravali Mobile Belts, and Chhotanagpur-Singhbhum Mobile Belt. The emplacement of granites in some mobile belts marked the end of the Proterozoic Era in the Indian Shield (Valdiya 2016).

The thick sedimentary successions of the Cuddapah and Godavari basins lie to the west of the Eastern Ghat Mobile Belt, and the Vindhyan Basin lies within the arm of the Satpura-Aravali Mobile Belts. The 900 to 750 Ma magmatic activities in Aravali are represented by Erinpura Granite and the acid Malani Rhyolite.

The coal-bearing Gondwana basins developed due to rifting of the Gondwana landmass during early Permian. The late Cretaceous outburst of basaltic lava in central and western India constitute the Deccan Trap Province with its intra-and inter-trappean sedimentary beds. The Upper Jurassic to the Quaternary pericratonic sedimentary basins from Kachchh in the west, Malabar and Coromandal coasts in the south, to Orissa in the east evolved due to faulting of the coastal belts.

The northern rampart of the Indian continent. The dynamic Himalaya is the product of the docking of India with Asia sometime 65–55 Ma ago. The pre-collision metamorphics in Lesser and Higher Himalaya and syn-to post-collisional sequences of Tethyan Himalaya with Indus Flysch form the complete gamut of the Himalayan belt. The Himalayan foredeep is occupied by the thick fluvial deposit of the Ganga river system drawn largely from the erosion of the rising frontal mountain belt (Valdiya 2016).

The chain of islands between the Bay of Bengal and the Andaman Sea makes the Andaman Island Arc, which constitutes the central part of the Indonesia-Myanmar Mobile Belt. These islands are made up of a succession of Cretaceous to Tertiary sedimentary rocks. The evolutions of these islands are related to the continued underthrusting.

3 Classification of Geoheritage Sites

India is blessed with cultural and historical diversity, however, geological heritage is still a lesser known term among the planners and policymakers. The wealth of geological heritage sites are scattered in different geological terranes and can be classified as below:

3.1.1. Fossil Parks (Dinosaur Fossil Parks, Siwalik Fossil Park, Fossil Wood Parks, etc.)

3.1.2. Metallurgical and Mineral-Rich Sites (Zawar in Rajasthan, Kolar gold mines etc.)

3.1.3. Sites of geomorphic Features (Columnar joints, Rock gardens, Caves, Impact structures, etc.)

3.1.4. Panoramic View Sites (Suture Zone at Ladakh, Deccan Trap sites etc.)

3.1.5. Sites of Geoscientific Significance (Nepheline Syenite, Welded Tuff, Barite deposit, etc.)

3.1 Fossil Sites

Fossils are impressions or moulds of animals and plants in rocks indicating the age of the rocks, life, and environment during their formation. Fossils are also living record of the history of evolution of life vis-à-vis geological time scales. As such, fossils largely attract the attention of not only geologists but also common people. Because of the attraction towards fossils, most of the fossil sites are well-identified in India and some sites are well known tourist destination. The Late Cretaceous dinosaur egg and bone fossils from Raiyoli (Gujarat) were discovered by the Geological Survey of India (GSI) and this news was widely publicised. The site remained unattended for a long time resulting into unscrupulous digging and selling of dinosaur eggs. Site could be declared protected only in 1997 by Government of Gujarat under *Bombay Police Act, 1951*. Later, a field museum was constructed which was recently renovated into a world-class museum and fossil park by the Government of Gujarat. Another significant Late Cretaceous dinosaur fossil site at Bagh (Madhya Pradesh) along the Narmada river valley, is of great geological importance due to the presence of intra-trappean undisturbed sedimentary sequence of Bagh and Lameta Groups representing inland marine transgressive-regressive sequence followed by dinosaur fossil bearing continental sediments. A National Dinosaur Fossil Park was declared by Government of Madhya Pradesh in 2011, but unfortunately no conservation and development activity took place. Only local conservation activists are taking care of these fossils.

The vertebrate fossils of Siwalik Group (Miocene-Pleistocene) preserved in the Himalayan foot-hill belt (*e.g.* hippopotamus, large tusked elephants, giant turtles, four-horned giraffe, saber-toothed cat, etc.). The GSI, in association with the Government of Himachal Pradesh, established the Saketi Fossil Park and Museum in 1974 to preserve these fossil sites.

Another National Fossil Wood Park, Tiruvakkarai (Tamil Nadu) conserves about 20 million years old fossil woods. The park was established way back in 1940 is well maintained by GSI (Fig. 2). These petrified tree woods indicate a forest cover during that era, which invites a lot of attention to palaeoclimatologists, palaeobotanists, and common citizens as it preserves the palaeo-forest records.

Precambrian stromatolites are recorded at different stratigraphic levels at several locations across India and are well studied. Many sites are though developed as Parks but either not properly maintained or not scientifically established. Such

Fig. 2 Fossil wood of Tiruvakkarai, Tamil Nadu (Photo courtesy: V. Chandramauli)

a. b.

Fig. 3 (**a**) Stromatolites of Jankikund, Chitrakoot, Uttar Pradesh (Photo courtesy: Mukund Sharma); (**b**) Bhojunda Stromatolite, Chittaurgarh, Rajasthan (Photo courtesy: Narendra Joshi)

stromatolite fossil parks are located in Uttar Pradesh and Rajasthan (Fig. 3a & b). However, rare Archaean stromatolites in Karnataka are left unattended.

3.2 *Metallurgical and Mineral-Rich Sites*

World's oldest mining and metallurgical records are well-known from India, which make us proud about the knowledge of our ancestors. The world's oldest metallurgical records of gold, lead, and zinc (about 2500 years old) scattered in Rajasthan and Andhra Pradesh are still not preserved. The American Society of Metals recognized ancient lead, zinc, silver mining, and metallurgy of Zawar (Fig. 4), Rajasthan in

Fig. 4 Retorts of zinc smelting in Zawar, Udaipur, Rajasthan used in a wall

1988. The unique metallurgical processing of ores to extract zinc which helped Europe in its industrial revolution. The ancient archaeological sites, remains of metallurgical processes, and ancient mining of lead and zinc constitute the area a fit case of UNESCO Global Geopark to highlight our heritage to the rest of the world.

3.3 Sites of Geomorphic Features

Surface physical features on the Earth carved out due to various geological, erosional, and chemical processes are another group of geoheritage sites, which attract people at large. The surface morphology shaped Earth due to combination of many geological processes are also sites of many cultural and historical heritage. Balancing rocks; river gorges, caves, columnar joints, erosional landforms, impact structures, and volcanoes are such important geomorphic features (Fig. 5). These sites educate visitors about the erosional process which are still operative. Some of these sites, though developed by different agencies, mostly lack the correct geoscientific background and knowledge to impress upon the visitors. However, even developing agencies are unaware of the significance and science behind such features. For example, Orvakal Rock Garden, Kurnool, Andhra Pradesh is one of the best-developed sites but unfortunately 'Rock Garden' does not provide any details about the rocks. Similarly, many caves are protected, mainly due to their spiritual value, but do not elaborate the geological processes involved in their formation. Many caves even display wrong information. Andaman Islands shows only active volcano and mud volcano of India (Fig. 5) and are one of the tourist sites.

Fig. 5 Active Volcano, Barren Islands (Photo courtesy: Alok Kumar)

3.4 Panoramic View Sites

Panoramic viewpoints are important sites in many hilly tourist destination. Many visitors appreciate scenic beauty but most of them do not understand what actually they are seeing. These sites are not subscribed with scientific knowledge information boards. For example, students are reading a lot about the collision of the Indian plate with the Eurasian plate and the evolution of the rising mighty Himalayas. But where are these exotic locations to witness with one's own eyes, is unknown. Shyok River of Ladakh cold desert forms the contact zone (suture zone) and there are several viewpoints on the bank of the river, but lack a geoscientific approach. Similarly, the Deccan Trap lava flows from Mahabaleshwar in the Western Ghats present exotic scenery but visitors do not understand it due to a lack of any onsite information (Fig. 6.). Interestingly, these are the least expensive geosites/geopoints capable of transferring knowledge to the next generation if properly explained. These sites are legislated as 'Areas of Outstanding Natural Beauty (AONB)' in European countries, and are also included within the ambit of designated geoparks. However, India does not have any such practice in tourist destination.

3.5 Sites of Geoscientific Significance

There are several geoheritage sites, which are textbooks of geological processes and extensively studied on national and global level. However, they may or may not attract the attention of the common people as a result they are least protected despite

Fig. 6 Deccan Trap lava flows of Mahabaleshwar, Maharashtra (Photo courtesy: Shyam Mude)

their high geoscientific value. So far, India has not passed any law to protect such scientific properties of the Earth and are in the state of neglect. There is an enormous threat to these important geological sections and treasures that are significant for future research to extract scientific information on Earth's processes. Geoscientists have the responsibility to raise their voice to save destruction of these sites from various developmental activities and ignorance about the importance of these sections. If known to planners and declared protected, the development can go hand in hand with the protection of such geosites. This is to emphasise that once any natural section gets destroyed, it is nearly impossible to reconstruct. Some examples are highlighted here to show the importance of some of the sections vulnerable to destruction. An approximately 250 million years old geological section in Spiti (Himachal Pradesh) records the event of the most disastrous mass extension in the Earth's history- the Permian-Triassic mass extinction. This section is at the roadside and under threat of being demolished very soon. The lignite deposit (about 56 million years old) of Gujarat yields some of the internationally acclaimed vertebrate fossils and the most interesting ancient 'amber'. Though mining can not be stopped because of the economic value of the minerals, some of the important geological sections should be easily preserved in consultation with geoscientists for future research. The Cretaceous intertrappean Anjar section in Gujarat, which contain the event bed, which is responsible for the extinction of dinosaurs, is almost destroyed. The largest barite mine of India is situated at Mangampeta, Cuddapah District of Andhra Pradesh. A geological section within the mine is also declared as a National Geological Monument. It is taken care neither by mining company nor by GSI. There are several such examples where geological wealth is at stake and needs to be take care by policymakers. The sustainable developmental plans can only save these natural treasures.

Recognition and Status

The criteria for recognition of geoheritage sites of International and National level are well defined by various agencies and we have to understand that every geological section can not be conserved and protected. So there is need to establish norms to classify geoheritage sites in India. The sites may be classified as International, National, or Regional and their scientific and/or geotourism importance should be identified. This is also to understand that all geoheritage sites can not be converted to a Geoparks and may or may not have tourism values but still they need to be protected and conserved for future generations. Many of the UNESCO World Heritage sites and tourist destinations in India are associated with interesting geological features but they neither been properly documented nor is known to developmental agencies. The Western Ghats, also known as the Sahyadri Mountain range, is a mountain range that extends about 1600 km parallel to the western coast of India. It is a UNESCO World Heritage Site and one of the 36 biodiversity hotspots in the world, which contains a very large proportion of the country's endemic flora and fauna. The fact is that this biodiversity is due to the geological diversity and the long escarpment is formed due to the coastal fault. The Deccan Trap lava flows and several geological sections are of international significance like Mahabaleshwar, which has been recognised under Second 100 IUGS Geological Heritage Sites (Fig. 6).

There are a large number of designated geoheritage sites as National Geological Monuments or Fossil Parks but very few are properly conserved and developed. The Cretaceous dinosaur egg fossils discovered from the Bagh area of Dhar District (M.P.) are under threat. The wood fossils of the dinosaurian period (about 66 million years old) were sold and destroyed so much so that it is now difficult to find a fossil wood sample in the locality. The Langza village of Spiti Himalaya which is also popularly known as 'fossil village', is a well known destination to purchase ammonite fossils of Triassic and Jurassic periods. Similar unlawful sale of 'geowealth' can be seen at different sites in India.

3.6 *Unofficial Register of Geoheritage Sites*

The importance of geological heritage was understood in India long before the global concept and Geological Survey of India (GSI) has declared a total of 32 geologically unique sites as 'National Geological Monuments (NGMs) over the years. These NGMs are spread all over the country mainly located in Rajasthan, Odisha, Karnataka, Andhra Pradesh, and Tamil Nadu. The declaration of NGMs started back in the 1940s and maximum NGMs were declared during the 1970s but while India was leading this concept, it could not keep pace with International developments due to shifting priorities. GSI recently provided a list of 90 National Geoheritage Sites available on its portal (https://bhukosh.gsi.gov.in/Geotourism). However, this list is incomplete and several other important geoheritage Hot Spots of India, which are of global significance, are not listed. The Society of Earth

Scientists prepared an 'Unofficial Register of Geoheritage Sites of India (URGI)' for the first time, which is prepared, based on available data. The geoheritage information along with URGI is available on website www.geoheritage.in. However, the majority of sites need to be properly documented and assessed for their geoheritage value.

3.7 Legislative Protection

There is no legislative protection for geoheritage wealth of India and because of this fact the Raiyoli Dinosaur site was protected under the 'Bombay Police Act, of 1951'. This is the only site in India declared protected lawfully under legislation cover. In the absence of any legislation, appropriate conservation and protection measures could not be taken. In 2019, The Society of Earth Scientists (SES) constituted a committee to prepare a status report and draft legislation, which was submitted to the government. Though, some deliberations were made, the legislation yet to come up. The awareness about geoheritage conservation is very poor among policymakers and political leaders. SES continued its efforts on different platforms to apprise them of the need to conserve geological heritage. As there is no dedicated organisation, SES proposed establishing a dedicated National Geoheritage Authority supported by State Geoheritage Authorities as there is no proper geoheritage regulating agency in India. There is an urgent need to bring legislation for geoheritage conservation in India.

4 Geotourism Potential and Geoparks

India is well known for its historical and cultural tourism. There are 42 UNESCO World Heritage Sites in the year 2024 among which, 34 are cultural, 7 are natural, and 1 is of mixed type. UNESCO World Heritage Sites- Western Ghats (Maharashtra); Hampi (Karnataka); Ajanta & Alora caves (Maharashtra); Sunderbans (West Bengal); Great Himalayan National Park (Himachal Pradesh); Bhimbetica (Madhya Pradesh) and Kanchenjunga National Park (Sikkim) had large potential to highlight associated geological heritage as a criteria to get UNESCO tag. But unfortunately their important geological features were not considered. As an example the balancing rocks and boulders of Hampi are very famous and could have been promoted along with World Heritage Site.

Unfortunately, Geotourism is still an unknown term even for tourism professionals in India. As a result, there is no Geopark even on a national level. There is a complete lack of popular literature on geoheritage and geotourism. The research papers published by geologists on geoheritage are full of critical geological terminology and scientific details. Therefore, they are less appreciated and have less impact on planners and policymakers. UNESCO is very clear on this point: 'A UNESCO Global Geopark must take great care not to alienate the public from science and must avoid the use of technical scientific language on information boards, signs, leaflets, maps and books which are aimed at the general public' (UNESCO 2016), which is also true in case of geoconservation and promotion of geotourism. The Society of Earth Scientists for the last few years has been approaching state government authorities for geoheritage conservation and geotourism through the development of Geoparks. In India, there are several geoheritage regions, which are fit to the norms of a Geopark but are neither developed nor recognised as Geopark due to a lack of understanding of this concept. We are supportive of establishing Geoparks than World Heritage Sites because Geoparks fulfill UNESCO Sustainable Development Goals; amalgamate culture, history, archaeology, biodiversity, etc. along with the geological heritage, and involve society at large. There are several sites, which are near to the definition of a Geopark but have not been properly documented and recognised. The Society of Earth Scientists has identified a few such sites and proposed to government agencies. For example; Ramgarh Crater in Rajasthan is the best-exposed site and has also been scientifically extensively studied. Ramgarh Crater is a well-known physiographic landmark in almost a flat sedimentary terrain. The 3.5 km diameter annular feature rises some 200 m above the surrounding flat plain (Kenkmann et al. 2019).

Chitrakoot region is geographically spread in the states of Uttar Pradesh and Madhya Pradesh. It has several internationally important cultural, geological, and geomorphological sites and is the most suitable region to be developed and promoted as a UNESCO Global Geopark. Shitole et al. (2023) assessed the potentiality of Bagh Dinosaur National Fossil Park region as a potential UNESCO Global Geopark and identified and documented geoheritage, historical and cultural sites. Bagh region has mainly a tribal population and their culture, traditions, and handicrafts (e.g. Bagh Prints) are points of attraction. Bagh caves are a group of nine rock-cut Buddhist caves belonging to the fifth and seventh Centuries AD. Only four caves are better preserved and are of archaeological importance. Therefore, it has all the ingredients to be promoted as a Geopark. **Gandikota**, the village on the right bank of River Pennar near Jammalamadugu, is located at a distance of 90 km from Kadapa, Andhra Pradesh and popularly known as the **'Grand Canyon of India'**. The huge fort, stunning gorge, lakes, temples, beautiful landscape, and wild forests bring it on the platform of a Geopark.

The knowledge of geotourism has an important bearing in enlightening society about sustainable environmental management and development, and when integrated with the cultural ecosystem and biodiversity helps people to reconnect with the past. The geotourism brings tangible benefits in the form of economic returns to local communities. Some of the lesser known geoheritage sites have been covered in this book Author/s have described in more of a popular fashion than purely in scientific way.

Mishra et al. described the world's most unique and only lithified parabolic dunes formed along southern coastal Saurashtra. These are constituted of fragmentary mollusk shells and coral reefs dissolved in calcium carbonate. The paper well illustrates the various sedimentary depositional features. These sites have never been considered for their geoheritage value.

Ladakh, in Trans-Himalaya, is a well-known tourist destination and exotic geoheritage sites. But these sites have not been properly documented and developed. Lal and Singh described geotourism potential of lakes and palaeolakes of Ladakh.

Misra and Misra described physical volcanic features associated with lava tube, channels, tunnels and effusive centers which have been utilized by the ancient artisans for creating some of the magnificent geoheritage sites in the Deccan region of Western India.

Saini et al. described geoarchaeological sites of the paleo Saraswati River in northwestern India, which are of popular significance proving presence of sacred Saraswati River. Kumar et al. illustrated geoheritage sites of Ladakh batholith, which have great record of evolution of granitoids and have research value. Pande et al. explained a case study of Patiya village cluster in Almora district of Uttarakhand where a Geotrail has been established showing geology of Almora nappe. Bundelkhand craton contain some of the oldest rocks on Earth and spectacular geoheritage sites may be developed to conserve the outcrops. Malviya et al. elaborated exotic geoheritage sites of this region. Deccan Trap Province (DVP) cover about 5 lakh sq. km. area in central and western India. It is one of the Largest Volcanic Province of the world. Kale elaborated exotic geoheritage sites of DVP including multiple types of columnar joins. Bharti has described geoheritage sites developed due to first marine transgression during Permian. Coprolites as a geoheritage sites is the new idea in this field. Kapoor explained why study and conservation of Coprolites is significant and their conservation will keep the record of dietary habits of animals. Tripathi et al. evaluated Chitrakoot region falling in Uttar Pradesh and Madhya Pradesh as potential UNESCO Global Geopark. The assessment concludes that the region is close to the definition of a Geopark.

The 'Hot Desert of Western India' offers some of the most spectacular geoheritages sites described by Wadhavan.

The book presents lesser known geoheritage sites of India which have not been documented so far. Hope it will add new dimension to the conservation of geoheritage sites of India.

Acknowledgements Authors are thankful to Narendra Joshi, Indore; V. Chandramauli, Chennai; Vishal Verma, Dhar; S.R. Jakhar, Jodhpur; Alok Kumar, Varanasi, and Shyam Mude, Pune for contributing good quality field photographs of geoheritage sites for this article. Authors are also thankful to Prof. Mukund Sharma, Lucknow for critically reviewing the manuscript.

References

Azzopardi E, Kenter JO, Young J, Leakey C, O'Connor S, Martino S, Flannery W, Sousa LP, Mylona D, Frangoudes K, Béguier I, Pafi M, da Silva AR, Ainscough J, Koutrakis M, da Silva MF, Pita C (2022) What are heritage values? Integrating natural and cultural heritage into environmental valuation. People Nat 5(2):368–383. https://doi.org/10.1002/pan3.10386

Gray M (2013) Geodiversity: valuing and conserving abiotic nature. John Wiley & Sons, p 512p

Gray M (2018) The confused position of the geosciences within the 'natural capital' and 'ecosystem services' approaches. Ecosyst Serv 34:106–112

Jade S, Shrungeshwara TS, Kumar K, Choudhury P, Dumka RK, Harsh B (2017) India plate angular velocity and contemporary deformation rates from continuous GPS measurements from 1996 to 2015. Sci Rep 7:11439. https://doi.org/10.1038/s41598-017-11697-w

Kenkmann T, Wulf G, Agarwal A, (2019) India's third impact crater: Ramgarh, Rajasthan. Large meteorite impacts VI 2019 (LPI Contrib. No. 2136) 5007.pdf

Molnar P, Tapponnier P (1977) The collision between India and Eurasia. Sci Amer 236(4):30–41

Shitole AD, Gupta V, Lata G, Verma V, Rajora S, Anand S, Tripathi SC (2023) Bagh dinosaur national park region, Dhar District, Madhya Pradesh: a potential UNESCO global Geopark. Geoheritage 15:116

UNESCO (2016) UNESCO Global Geoparks. https://unesdoc.unesco.org/ark:/48223/pf0000243650

Valdiya KS (2016) The making of India: geodynamic evolution. Springer, SES Series, p 924p

Unique Geoheritage Site of Lithified Quaternary Calcareous Parabolic Dunes and Their Preservation in Southern Saurashtra, India

Anshuman Misra, K. S. Misra, and Neela Misra

Abstract The world's most unique and only lithified parabolic dune field is best exposed in Una-Delvada-Diu region, along southern coastal Saurashtra. The individual dunes are composed of mechanically formed calcium carbonate sediments, formed due to the fragmentation of mollusk shells and coral reefs. Lithification of these well sorted and rounded paletoid sediments by dissolved calcium carbonate has provided very high chemical purity (98% $CaCO_3$). These lithified rocks are part of widespread Quaternary deposits commonly described as Miliolite limestone. The origin of these deposits has been debated in the past. One group believes they are of marine origin, while the others advanced an aeolian origin. The present study considers that all the sediments formed in marine, very high-energy beach environment. However, their deposition took place by both aeolian as well as marine processes. Parabolic dunes and pile-ups on the western and south-western sides of hills were deposited by aeolian processes, during the arid conditions, corresponding with glacial periods. During these periods, sea levels went down and exposed the beach sediments to be transported to interior regions by strong winds. In contrast the intervening humid interglacial periods were marked by the rise in sea levels, flooding of the coastal plains and valley portions, where the inter-bedded sequence of coarse and fine fractions were deposited under rhythmic, agitated and calmer conditions, along with mega fossils of gastropods, lamellibranches and echinoderms. Dominant prevalence of southwesterly winds is deduced during the entire Quaternary Era. It has vibrant white and yellow color, uniform texture, very suitable for intricate carving, easy quarrying along laminations and sawed to desired dimensions. These represent relatively rare, attractive and unique geological features. As deduced from the

A. Misra (✉)
Department of Geology, Doon University, Dehradun, Uttarakhand, India

K. S. Misra
Department of Petroleum Engineering and Earth Sciences, University of Petroleum and Energy Studies, Dehradun, Uttarakhand, India

N. Misra
Department of Geography, Savitribai Phule University, Pune, Maharashtra, India

S. C. Tripathi et al. (eds.), *Geoconservation and Geotourism Potential of India*, Society of Earth Scientists Series, https://doi.org/10.1007/978-3-031-81017-6_2

multi dated satellite data it is seen that the outcrops of the parabolic dunes are vanishing very fast in recent years due to indiscriminate quarrying as well to reclaim agricultural fields. There is an urgent need to identify an area where they can be best preserved as a geoheritage site for future generations. Furthermore, this pelletoid unit is also an excellent heritage stone, being used for centuries in construction of buildings not only in Saurashtra but also in far off places such as Mumbai and Yangon. Scope for linking this site with other geo-heritage locations in its vicinity is also discussed.

Keywords Quaternary calcareous deposits · Miliolite limestone · Parabolic dunes · Geoheritage sites and preservation · Climate and sea level changes · Wind directions · Saurashtra · India

1 Introduction

Quaternary calcareous deposits of Saurashtra in western India have been engaging the attention of geologists ever since Carter (1849) described and named them as Miliolite Limestone. Fedden (1884) gave detailed description of geology and stratigraphy of Saurashtra, earlier known as Kattyawar. Evans (1900) recognized that these limestone units have formed mechanically by the fragmentation of mollusk shells and coral reefs. In scientific literature, these deposited and lithified calcareous sediments are known as Miliolite limestone due to recognition of foraminifer Miliolina. However, subsequent detailed paleontological studies indicated that they are not only deficient in genus Miliolina but also other genus belonging to family Miliolidie, (Sastri and Pant 1959, 1960). Srivastava (1968) did detailed petrographic examination of these rocks and identified calcarenites, calcirudites, coquinoid bands, and minor carbonate-mud, formed under shallow, warm, agitated waters of tropical sea shores, deposited during higher stand of the sea during Pleistocene. Mathur (1978) described calcirudite unit as Chaya Formation. Pandey et al. (2007) emphasized the statigraphy distribution and depositation environment of the Chaya Formation of costal Saurashrta. Our studies indicate that this formation was deposited in low energy lagoon environment. Often they have complete fossilized shells of gastropods, lamellibranches and echinoderms. Calcirudite (Chaya unit) with large fossils of lamellibranch, gastropod and echinoderm have been certainly deposited by marine transgressions Since there are many lateral facie variations, it is difficult to identify as a mapable unit. With slight modification and suitability we included carbonate mud and coquinoid bands within the calcmicrite unit. Among these three sedimentary units, the calcarenite is most dominant (approximately 90%) and occurs both as aeolian dunes, pile-ups on the southwestern sides of hills, as well as the inter-bedded sequence of calcarenite and calcmicrite. Commercially the calcarenite unit is known as Porbandar limestone and is the most favored dimension stone in Saurashtra. It can be easily quarried due to the invariable presence of lamination, can be sawed to desired shape and fixed as large building blocks or slabs. Absolute white and vibrant willowish colors have made it the preferred dimension stone for magnificent churches in Diu and Somnath temple on the southern and western coast. Furthermore, it has the property of getting toughened by dissolved calcium carbonate subsequently. Since calcarenite unit comprises a large portion, and described as Miliolite limestone for more than one hundred and seventy years; the present authors prefer and suggest continuance with the same name.

The biggest controversy raged and continues till date concerns the origin of these deposits. One group advocated marine origin and suggested many evidences to support the view, while others believed in an aeolian origin. Biswas (1971) suggested aeolian origin and indicated southeasterly winds deposited them from southern Saurashtra right across in the Kutch region, Lele (1973) put forward marine origin. Misra and Misra (1982) studied the morphology of Quaternary deposits of entire western India as well as the parabolic dunes. They concluded that the dominant wind direction during entire Quaternary Era continued to be from south-west and corresponds with the prevailing one. Our field studies have shown that these deposits occur as over-burden on all older rock formations; south of Khakarathal, they overlie Gondwana sediments, all around in interior Saurashtra on Deccan volcanic rocks, along coastal zone on Gaj Beds of Lower Miocene as well as Dwarka Beds of Pliocene age. Although they are mainly confined to the coastal zone, in exceptional cases, they occur in interior places in the proximity of Bamanbore, approximately 150 Km. from the nearest coast.

Multi-date satellite images have been interpreted to study and ascertain dwindling number of parabolic dunes through the years. The earliest data interpreted are aerial photographs taken in 1964. With the availability of remote sensing data, Landsat images of 1984, 2002 and 2020 were interpreted to ascertain the dwindling pattern. This exercise has revealed that the number of parabolic dunes has reduced steadily till 2020, however, this has accelerated in following years and they have completely disappeared in certain areas, such as south of Una, due to urbanization. Only in highly rugged terrain they are present because they could not be easily converted to agricultural land.

2 Quaternary Deposits of Western India

Both sandy and calcareous sediments are present in western India. The sandy sediments are mostly confined within the Thar Desert of Rajasthan and Cambay graben in northern Gujarat, while calcareous sediments are dominantly confined to Saurashtra. Vast desert is located between the Indus River in the west and the Aravalli mountains in the east and comprises enormous accumulation of sand dunes and sheets. Sandy sediments are contributed from the Indus delta region, where they form longitudinal dune fields. These dunes have formed in the proximity of source sediments and by very severe winds. With increasing distance from source, the longitudinal dunes gradually diminish and barchans dunes start to dominate. The salt has come from the vast flat Rann of Kutch areas. These areas get inundated by sea water during monsoon on yearly basis and desiccation produces enormous amount of salt. Presently no salt is lifted by the southwesterly winds. However, during the arid cycles the winds were much more severe and they lifted and transported the salt to interior regions. The desert extends from north of Rann of Kutch for about 600 km in the north-easterly direction and has a width of about 160 km. There are pile ups of sediments on the southwestern faces of the older Aravalli Ranges. Sand sheets fill the valley portions and occupy the pediplains. Older and younger pile ups and sheets can be identified by coloration, changing from yellowish to brownish shades. The Cambay Graben region is unique and has very thick accumulation of Quaternary sediments. Drill holes for the exploration of hydrocarbons have logged up 400 m thick alluvium deposit along the central axis. The graben portion is dominated by elongated fore dunes trending along northwest-southeast aligned graben. Evidence suggest that the entire graben was getting inundated during the interglacial periods while during arid periods it got exposed and supplied sandy sediments for the parabolic dunes of northern Gujarat right up to Ahmedabad (Misra and Misra 1982).

3 Quaternary Calcareous Deposits of Saurashtra

Both marine as well as aeolian types of Quaternary calcareous deposits are best located in different parts of Saurashtra. In totality, these sediments have formed on vast submerged continental platform, west of Mumbai coast. It had suitable geomorphic and climatic conditions for the prolific growth of mollusks and corals. Global climatic change to glaciation, lowered the sea level and exposed the platform to a very high energy beach environment, where fragmentation and breaking of shells of mollusks and corals generated the calcareous sediments comprising approximately 90% well-rounded pallets and rest oolites and sand grains. The oolites have concentric growth of calcite and radial aragonite.

The marine deposits comprise horizontally disposed inter-bedded sequence of palletoid and calcmicrites (chalk) in vast plains between the coastal ridges and interior rugged volcanic terrain. These seem to have been deposited during high sea levels corresponding with the humid inter-glacial periods. Palletoid bands are deposited under agitated conditions, while calcmicrites were deposited due to settling of fine fraction during calmer conditions. The average thickness of this sequence is approximately fifty meters and it is also inter-layered with localized patches of Calcirudites. These sediments are neither well sorted nor rounded, deposited in low energy lagoon environment, have rusty reddish colors with large fossils of mollusks and echinoderms, confirming their deposition in marine environment. Aeolian deposits are mainly palletoid unit and are located on the western and southwestern slopes of older Deccan volcanic hills. Two of the best pile-ups are exposed in Adi Kadi Vav and Adityana quarry sections are illustrated here.

Adi Kadi Vav also known as Adi Chadi Vav is an excellent section located in Uparkot Fort, in close proximity of Junagadh city. This oldest step well presents rock cut section on the south western face of Girnar Volcanic complex (Fig. 1). It is a fascinating site for the tourists and also a religious place. From geological point of view this symmetrically cut section is very revealing and is perhaps the best example of aeolian deposits with elaborate cross lamination. The present

Fig. 1 The photograph illustrates cross laminations, sets can be seen continuing for more than 50 m in palletoid unit deposited by aeolian processes, exposed in Adi Kadi Vav step-well section at the base of Girnar Hill

Fig. 2 Photograph shows details of cross lamination in Adi Kadi Vav section. The corrugation is due to textural variations. Lamina with coarser fraction stands out while finer one forms depression

study also recommends it to be considered for protection as an important geoheritage sites.

Closer examination of cross lamination has revealed that lamina composed of coarser fractions are slightly harder and thus stands out in relief, while finer fraction gets dissolved and erodes fast to form depression (Fig. 2).

Adityana quarries, located on south west face of Alec and Barda hills close to Ranavav (Fig. 3). These quarries supply most of the requirements of the dimension stone and are one of the classic examples of aeolian deposits with elaborate cross lamination. Since these quarries are unique and have typical characters of aeolian deposits, at least one quarry needs to be preserved as a geoheritage site.

The dimension stone industry has reduced significantly due to avalibility of ceramic tyles and has reduced requirement, however, slabs of Porbander limestone are produced and utilized because of asthetic value (Fig. 4).

Fig. 3 Adityana quarry section of palletoid unit on south west face of Alec and Barda hills, close to Ranavav. Cross laminations developed due to aeolian deposition and there utilization in excavation can be seen

Fig. 4 Slabs of Palletoid limestone in final finished form, available in Saurashtra

Apart from lithified deposits there are other calcareous sediments as fore dune complexes in coastal region and beach sediments. They represent unconsolidated recent sediments, are characteristically similar to the bulk composition and texture of palletoid unit.

4 Lithified Parabolic Dunes

The lithified parabolic dunes in the vicinity of Una-Delvada-Diu region in southern coastal Saurashtra are composed of highly calcareous pallets formed in high energy environment from shell fragments of mollusks and corals, quite similar to the present day beach sediments. The parabolic dunes are large and more in number, near the coast and their number and size decreases in northeasterly direction. The horns of these dunes point in southwestern direction (Fig. 5), along with gentle windward sides. The cross laminations on steep leeward sides dip up to 38° in northeast direction. The wind direction deduced from their shape and attitude of cross laminations is from southwestern direction.

Formation of lithified parabolic dunes seems to be due to combination of several curious conditions. Winds laden with moisture are required for formation of parabolic dunes; otherwise if the winds are dry, barchans form. High humidity of the winds is due to the proximity with the ocean. However, severe winds are the additional requirement for lifting the sediments and carrying it to several kilometers. Near the source the parabolic dunes are larger and more in number and the number reduces and size decreases with distance (Fig. 6).

Fig. 5 Landsat satellite imagery of lithified parabolic dune field in Una-Delvada-Diu area along the coastal southern Saurashtra

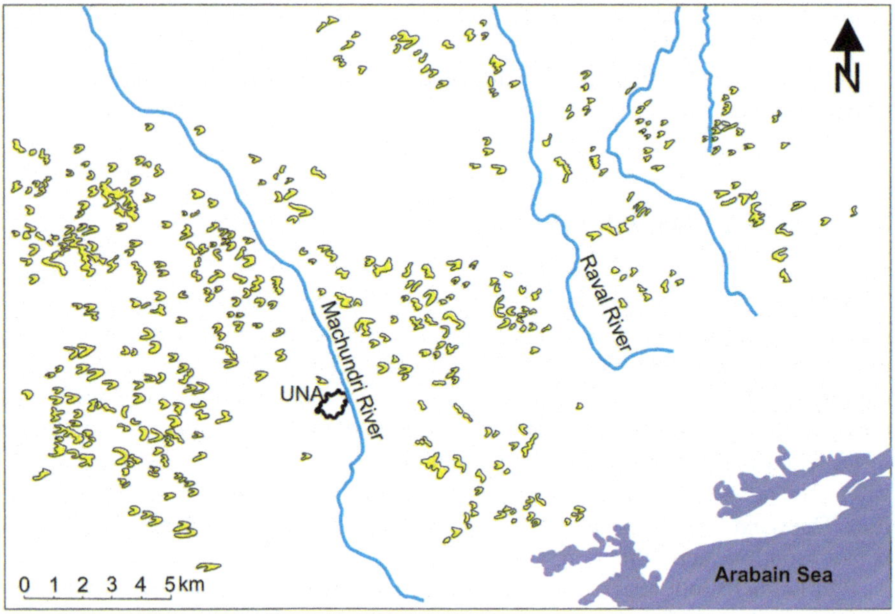

Fig. 6 Map shows distribution of lithified calcareous parabolic dunes in the vicinity of Una

5 Quaternary Climatic and Sea Level Changes

Various geomorphological evidences suggest that western India experienced repeated and severe arid and humid periods during Quaternary Era. The lithified parabolic dunes of palletoid limestone near Una-Delvada-Diu area in southern Saurashtra, extending up to 15 km inland from the present-day coast, thick accumulation of Aeolian palletoid units on the western and southwestern slopes of Girnar hill, large stabilized parabolic sand dunes in the vicinity of Ahmedabad, several generations of mega dunes and sand sheets in the Thar desert, point towards severe arid periods. On the other hand, very wide river courses of Sabarmati and Mahi, with extensive fluvial flood plains within small catchment area, all indicate very high rainfall during intervening humid periods.

Present study, has led the authors to believe that the entire western India witnessed a number of severe arid and humid cycles, corresponding with the glacial and inter-glacial periods of Quaternary Era. Innumerable geomorphological evidences are identified. During the arid (glacial) period, lot of water got locked in high latitude and altitude; the sea level went down and exposed vast beaches. Very severe winds transported the beach sediments and deposited it in interior regions. While during interglacial wet periods, sea levels have risen leading to in-ward shifting of beaches and deposition of inter-bedded, horizontally disposed sequence of palletoid and calcmicrite units in inner coastal plains. The accumulated dunes were consolidated and lithified during this period. The disposition of parabolic dunes suggests, that the southwesterly winds prevailed during their deposition also.

6 Quaternary Wind Direction

Biswas (1971) suggested that south-east winds deposited the Miliolite Limestone in Saurashtra and Kutch. His argument is based on eastern limit of outcrops up to Gopnath Point and transportation for hundreds of kilometers all across Saurashtra and Gulf of Kutch. Glennle (1970) observed in Abu Dhabi region that during lateral transportation from the common siliceous and calcareous source sediments, the sand content steadily increases while there is decrease in calcareous content. Absence of such a change and purity of calcareous sediments does not support the idea of large scale lateral transportation. He also concluded that in Rajasthan desert, major fore set slopes resulted from the southwest monsoon winds, but 2 months after the monsoon a milder wind blows from the north east. Lele (1973) concluded that from January to April and October to December low velocity northerly winds prevail, while westerly is the predominant direction for the months of May to September. Hegde (1977) studied parabolic sand dunes in northern Gujarat near Kalol and Palanpur and concluded southwesterly wind direction. The other region of stabilized parabolic dunes are in the vicinity of Ahmedabad are composed of fine sandy sediments. The wind direction deduced from their study, has also suggested southwesterly winds. Misra and Misra (1982) described the morphology of aeolian deposits of entire western India, including the parabolic dune of Saurashtra and concluded that dominantly southwesterly winds prevailed during the arid cycles of Quaternary Era.

7 Protection of Parabolic Dune Complex as Geo-Heritage Site

Urgent measures are required to protect the only and unique out crops of the parabolic dunes in the world. They are vanishing very fast due to rampant quarrying for dimension stone for construction of houses and other structures. Apart from this the dune areas are excavated to get plain agricultural fields (Fig. 7). Multi-date mapping done by interpreting aerial photographs (1:50,000 scale) taken in 1964 shows much higher number of dunes in the vicinity of Una-Delvada-Diu area. Later interpretations on the same scale by using Landsat satellite images of 1984, 2002 and 2020 show that the reduction in number of dunes reduced gradually, however, since 2020 the disappearance has significantly accelerated. It is utmost priority to notify an area, where they are best preserved with associated natural features with exposed sections of various sets of cross laminations. Preservation of dunes within it and ensure their continuity for future studies and the posterity. It is of utmost importance to identify suitable area where they are exposed in their original form and in association with each other. Importance will increase, preferably if a cross section of a dune is also exposed.

Fig. 7 It can be seen in this satellite picture that only very few parabolic dunes are remaining in landscape reclaimed for agriculture in recent years

Our study also recommends, three other magnificent geoheritage sites of great geological interest, associated with calcareous deposits, are Adityana quarries, Jambavan cave and Adi Kadi vav section. Adityana quarries for sedimentary cross lamination, Jambavan cave for inter-bedded sequence of calcarenite and calcmicrite units both in the vicinity of Ranavav near Porbandar, sediment pile up at Adi Kadi Vav section on the south-western side of Girnar Hill complex and caves in Diu Island. Jambavan cave has formed due to dissolution of calcmicrite unit by percolating water. Roof and floor are made up of coarse palletoid units. (Fig. 8). Furthermore, percolation water has formed both stalactite and stalagmites.

The other highly impressive cave within the calcareous sediments is located in Diu Island. This cave has formed due to dissolution of aeolian palletoid unit with multi directional top and bottom sets of cross lamination. Dissolution has reached to lower levels and large portion of the cave is filled with ground water. These nicely illuminated and maintained caves are invariably visited by the tourists of Diu Island (Fig. 9).

The parabolic dune and other recommended sites are well approachable from the scenic coastal highway, dotted with on and off coastal dunes, white calcareous beaches, clear sea water, branching and other type of palm trees along with modern tourist facilities are added attraction. Several geological sites such as Girnar Igneous Complex, Tulsi Shyam hot springs, a variety of dipping alkaline and felsic lava flows are located in contiguous areas. Uparkot Fort, Jama Masjid, Mahabat Maqbara, Buddhist caves and Mahatma Gandhi birth place at Porbandar are important historical sites in nearby areas. From religious point of view very attractive Somnath, Dwarka and Jain temples of Palitana and Girnar along with magnificent churches of Diu are not far. Furthermore, Asiatic lion sanctuary and zoological parks can also be added to tourist cluster.

Fig. 8 Photograph showing the main hall of the Jambavan cave close to Ranavav near Porbandar

Fig. 9 Cross lamination in aeolian palletoid unit exposed in natural cave section Diu Island. This is an important tourist site and proper illumination has enhanced the features

8 Summary and Discussions

Evidences suggest that the parabolic dunes have formed in two stages. During the first accumulation of sediments has taken place by severe locally moisture laden winds in an otherwise arid period. Lowering of sea level has exposed vast beaches to a phase of wind transportation and accumulation. Followed by second stage of humid rainy inter-glacial period, where litification has taken place by tepid rain water. The study of snow dunes in eastern Antarctica (Misra et al. 2023) has also suggested formation in two stages. The snow was accumulated as long linear ridges corresponding with the fore-dunes. They are also perpendicular to the wind direction with gentler wind ward and steeper lee ward sides. The contrasting difference, however, is that the deposition of snow dunes has taken place during inter-glacial blizzard conditions while the compaction has taken place during dry and very cold glacial period.

The present study removes several misconceptions. It has been made clear beyond doubt that the calcareous sediments have formed by mechanical fragmentation of shells of lamellibranch, gastropod and coral reefs in high energy beach environment. The three sedimentary units, palletoid (Miliolite) limestone, comprise approximately 90% well-rounded and sorted pallets, 5% oolites and rest sand grains. Horizontally disposed inter-bedded units of this along with fine calcmicrite (chalk) and low energy lagoon deposits calcirudite (Chaya unit) with large fossils of lamellibranch, gastropod and echinoderm have been certainly deposited by marine transgressions. The arid cycles corresponded with glacial periods of high latitude and altitude, while humid periods were associated with wet interglacial periods of Quaternary Era. Arid periods were also characterized by scanty rainfall, very severe and dry winds. Most of the dunes and pile ups have formed during these periods. The inter-glacial periods were hot and humid, very high rainfall accompanied with rise in sea levels and deposition of inter-bedded sequence of palletoid and calcmicrites in marine conditions. Palletoid units were deposited during agitated conditions while calcmicrite formed by settling of fine fraction in calmer conditions. Calcirudites were deposited in low energy lagoon conditions. Furthermore, parabolic dunes and pile ups of palletoid units were lithified during these interglacial periods.

This study also addresses the problem of rise and fall of sea levels during Quaternary Era. Deposition of marine sequence up to approximately 70 m elevation, suggests that maximum rise in sea level reached at least to this level. Large portions of Saurashtra got submerged below this level; new beaches were created and in turn contributed to accumulations at higher levels in central and other parts such as Bamanbore near Chotila Hill. Mathur et al. (2004) concluded that major sea level changes were during Pleistocene, while they were quite minor 2 to 4 m during Holocene. Prolific growth of mollusks and corals in warm tropical sea, fragmentation by severe agitated conditions formed well sorted and rounded calcareous sediments. Transportation has taken place by strong winds and deposited due to the changing conditions as parabolic and other forms of dunes, along with pile ups on southwestern slopes also composed of palletoid unit. Locally moisture laden winds in an otherwise severe aridity have deposited the parabolic dunes.

Because of their very high purity, the coarse fractions are used as dimension stones as well as in chemical and cement industry, while chalk is extracted from the fine fraction. Quaternary deposits of Saurashtra host some of the finest geoheritage sites, over exploited and need very urgent conservation. The important ones are lithified calcareous parabolic dunes, enormous pile up on western faces of hills characterized by long cross laminations, exposed in rock-cut sections of huge step-well (Adi Kadi Vav) and extensive cave (Jambavan), developed due to dissolution of finer fractions.

9 Conclusions

A series of rather unique geological setting of vast and shallow continental shelf, synchronized with Quaternary climatic changes accompanied by sea level changes, have generated and transported the sediments to form the parabolic dunes. It has also been concluded that dominantly southwesterly winds prevailed during the Quaternary Era. During arid periods, the severity was so high that it has lifted palletoid sediments, carried to several kilometers and deposited both as dunes as well as pile ups. Present day seasonal changes are of significantly low intensity in comparison to the arid and humid cycles of Quaternary. The winter and summer together correspond with arid while rainy season represents humid cycle. During Quaternary the severity of both arid and humid periods were much more pronounced. Sea levels have also oscillated within larger range. Precipitation was also significantly heavy during humid inter-glacial periods. All the geomorphic evidences suggest repeated climatic changes. Presently we are living in inter-glacial period, however, whether we are approaching a more intense inter-glaciation or a new cycle of glaciation is setting in, is not yet resolved. The subtle evidence to make any suggestion are masked by repeated major changes.

Parabolic dune fields are composed of high purity (98%) calcium carbonate sediments that have formed in high-energy beach environment and deposited by the aeolian processes. The disposition of parabolic dunes suggests dominantly southwesterly winds prevailed during their formation. The very shape of the Thar desert comprising largest accumulation of blown sands. Longitudinal dunes trending in SW-NE direction north of Rann of Kutch, formation of large sand-sheets on the western sides of mountain ranges, gentler SW windward and steeper NE lee ward sides of transverse dunes, and stabilized parabolic dunes in the vicinity of Ahmedabad, indicate that dominantly SW winds prevailed during the Quaternary Era. This is in agreement with the present day wind direction. The important fact thus emerges out is that there is no major change in the wind direction in western India since the beginning of Quaternary.

The parabolic dunes are disappearing very fast because of quarrying and reclamation of land for agriculture. The study has revealed that the rate of disappearance has got accelerated since 2020. It is urgently required to save them as a protected geoheritage site. The site is located along scenic coastal highway, with unique branching palm trees on either side. Furthermore, it is closer to innumerable religious, historical, geological and Asiatic lion sanctuary and this unique geoheritage site can also have added touristic attractions.

The parabolic dune and other recommended sites are well approachable from the scenic coastal highway, dotted with on and off coastal dunes, white calcareous beaches, clear sea water, branching and other type of palm trees along with modern tourist facilities are added attraction. Several geological sites such as Girnar Igneous Complex, Tulsi Shyam hot springs, a variety of dipping alkaline and felsic lava flows are located in contiguous areas. Uparkot Fort, Jama Masjid, Mahabat Maqbara, Buddhist caves and Mahatma Gandhi birth place at Porbandar are important historical sites in nearby areas. From religious point of view very attractive Somnath, Dwarka and Jain temples of Palitana and Girnar along with magnificent churches of Diu are not far. Furthermore, Asiatic lion sanctuary and zoological parks can also be added to tourist cluster.

10 Data Availability

All data or data source incorporated are mentioned in the article.

Acknowledgement We wish to profusely thank Dr. Satish C. Tripathi, Formerly Deputy Director General, Geological Survey of India; Honorary Editor, Earth Science India and General Secretary, The Society of Earth Scientists; for inviting us to contribute a chapter to forthcoming volume on Geoheritage. Landsat images from Google Earth, used during the study are thankfully acknowledged. Field guidance and innumerable discussion sessions with Dr.U B Mathur,Deputy Director General, Geological Survey of India, inculcated interest on this subjact. Valuable suggestion by Prof. N C Pant, Geology Department, Delhi University have helped to improve the manuscript are also thankfully acknowledgeed. Field trip with late Prof. Nilesh Bhatt of Geology Department, M. S. University of Baroda, in preparation of Field Guide for 36th International Geological Congress, motivated us to this write up on data collected earlier.

Funding No funding is received to do this research.

Ethics Declarations Competing Interests

The authors declare no competing interests. There are no financial or non-financial interests that are directly or indirectly related to the work submitted for publication.

References

Biswas SK (1971) The Miliolite rocks of Kutch and Kathiawar, Western India. Sediment Geol 5:147–164

Carter HJ (1849) On foraminifera, their organization and their existence in fossilized state in Arabia, Sindh, Kutch and Khattyawar. J Roy As Soc Bombay 3:158–173

Evans JW (1900) Mechanically formed limestones from Junagarh (Kathiawar) and other localities. Quart J Geol Soc London 56:559–583

Fedden F (1884) Geology of Kattyawar peninsula in Guzerat. Mem G.S.I. 21

Glennle KW (1970) Desert sedimentary environments. In: Developments in sedimentology, vol 14. Elsevier Publ. Co., Amterdam, p 222

Hegde KTM (1977) In the late quaternary environments in Gujarat and Rajasthan. In: Agrawal DP, Pande BM (eds) Ecology and archaeology of W. India: 217¬24. Concept Publication, Delhi

Lele US (1973) The Miliolite limestone of Saurashtra, Western India. Sediment Geol 10:303–310

Mathur UB (1978) Studies of Quaternary carbonates of southern Saurashtra with special reference to the Miliolite problem. Ph.D. Thesis, M.S. University. Baroda (unpublished).

Mathur UB, Pandey DK, Bahadur T (2004) Falling late Holocene Sea-level along the Indian coast. Curr Sci 87(4):25

Misra N, Misra KS (1982) Morphology of Aeolian deposits of Western India and their bearing on wind direction during quaternary era. In: Merh SS (ed) Proceedings first National Seminar on quaternary environments. Hindustan Publishing Corporation (I), Delhi

Misra A, Misra KS, Dobhal DP (2023) Compacted snow dune complexes in Antarctica and their applicability as new climate change and basement tectonic parameters. In: Phartiyaletal B (ed) Limate change and environmental impacts:past, present and future perspective. Society of Earth Scientists Series. https://doi.org/10.1007/978-3-031-13119-6_12

Pandey DK, Bahadur T, Mathur UB (2007) Stratigraphic distribution and depositional environment of the Chaya formation along the northwestern coast of Saurashtra peninsula, Western India. J Geol Soc India, V 69(June):1215–1230

Sastri VV, Pant SC (1959) Survey of coastal areas for collecting specimen of beach rock, Miliolite etc, for study of change of sea level. Geol Surv India. (Unpublished Report)

Sastri VV, Pant SC (1960) Foraminifera from the Miliolite of Saurashtra. Dent Abstr. 47th Indian Sci. Congress

Srivastava PK (1968) Petrography and origin of Miliolite limestone of Western Saurashtra coast. J Geol Soc India 9(1):88–96

Lakes and Palaeolakes of Ladakh in Context of Geoheritage and Geotourism

Ravish Lal and Preeti Singh

Abstract Ladakh, 'the Cold Desert' in the NW Himalayan and Trans-Himalayan zone represents an important geological and geoheritage site. The region holds significant sites related to understanding of present and past climatic changes along with hidden secrets of the India-Eurasia collision. Glacio-fluvial, fluvio-lacustrine and lacustrine preserves reliable indicator of causative processes, past climate and tectonics of the area. One of the most prominent geological features of this region is the presence of many lakes, generally, formed as a result of reactivation of faults or mass movement and are widespread all along the Ladakh region (present day lakes as well as paleolakes). These lakes are relatable to the fluctuation in glaciations and interglacial periods in the geological past (~Quaternary period). In addition, they are related to the neotectonic movement along the Karakoram fault (a major crustal scale fault) that further played a vital role in drainage disruption in the region as well as complex geomorphic processes forming variety of intertwined landforms.

We consider this area to have an incredible potential for developing as geoheritage and geotourism site. These Geo-sites will not only provide tourist attractions but also give an insight about extreme geological processes.

Keywords Ladakh · Geoheritage and Geotourism · Lakes and Paleolakes

---

R. Lal (✉)
Ram Lal Anand College, University of Delhi, Delhi, India
e-mail: ravishlal.geology@rla.du.ac.in

P. Singh
Department of Geology, University of Delhi, Delhi, India

S. C. Tripathi et al. (eds.), *Geoconservation and Geotourism Potential of India*, Society of Earth Scientists Series, https://doi.org/10.1007/978-3-031-81017-6_3

1 Introduction

The Indian subcontinent is endowed with diverse sites and archives of geoscientific significance, which also has a popular context and represents potential geotourist areas. The unique terrain of the union territory of Ladakh represents one such domain. Ladakh ("La" means passes and "Dhak" means numerous) is considered to be the home of snow- capped mountains encompassing one of the world largest non polar glaciers (e.g., Siachin Glacier; Umling La /5797 m, Khardung La /5600 m, Taglang La Pass/5328, Zoji La/3528 m above mean sea level). In general, it is a land blessed by nature treasure marked by amazing landscape beauty relatively unspoilt by man and cultural heritage marked by royal palaces and colourful monasteries (Fig. 1a and b). Ladakh is a home of the tribes of Tibetan stock of the Mongoloid race. Here, Buddhist culture is common and it is a common sight to see monasteries, locally named Gompa, decorated with prayer flags.

This terrain has rugged topography having barren mountains alongwith deep gorges, wide river valleys with little or no large-scale vegetation (Fig. 1c). The geo-scientific significance of this domain stems from both the geological and geomorphological perspective as tectonically this area is considered as an active highest plateau on the planet earth that is uniquely located in low latitudes and higher altitude (average elevation>3000 m amsl), covering an area of approximately 98,000 sq. km of the northwest Tethyan Himalayas and Trans-Himalayas. The northward movement of India progressively narrowed down the Tethyan Ocean (early Cretaceous-Lower Eocene) and was accompanied by subduction of the ocean floor under the crust of Eurasia (evidenced as Dras-Shyok volcanic island arc). The subduction produced a large amount of magmatic rocks and the magma formed the mountain range known as Trans-Himalayas on to the southern margin of the Eurasia and intra oceanic subduction within the Neo-Tethys followed by continent-continent collision leading Himalayan orogeny (Dewey and Burke 1973; Le Fort 1975; Searle et al. 1988, 1998; Searle 2015; Kohn 2014). The Himalayan mountain ranges are bordered in the west and north-west side by the mountain ranges of Karakoram Mountains, Hindu Kush and Pamirs ranges and in the north of the Himalayas lies Tibetan Plateau. The Trans-Himalayas Mountain Region or Tibet Himalayan Region is located to the north of the Great Himalayas which consists of Karakoram, Ladakh, Zanskar and Kailash mountain ranges (Fig. 2). The Karakoram Range is known as the 'backbone of high Asia' and is considered as an important link between Pamir and Himalaya.

Ladakh preserve events of our earth's history related to the reconstruction of the Indian Contineral Crust that give rise to one of the mightiest and magnificient mountain ranges in the world. General consensus is that during the evolution of Himalayas the first mountain range to emerge was the Trans-Himalayan igneous arc and Indus-Tsangpo were the first rivers to arise and transport sediments (during Eocene time). Basins in the Kargil, Kailash and Lhasa are filled with continental sediments brought by these rivers (Fig. 2). Two main mountain ranges in the Ladakh include Zanskar and Karakoram. Broadly, the area is bounded by the Indus-Tsangpo Suture Zone (ITSZ) in the south and the Shyok Suture zone (SSZ) in the north and is pivotal for

Fig. 1 Rich cultural heritage and architectural marvels of Ladakh, famous for colourful monasteries, royal palaces and beautiful landscape. (**a**) Leh Palace, former royal palace also known as Lachen Palkar Palace in Leh town in Ladakh. (**b**) One of the oldest and largest Buddhist monasteries, Diskit Monastery also known as Diskit Gompa located in the Nubra Valley of Ladakh, showcasing a large statue of Crowned Budha situated on a hil. (**c**) View of the Indus river flowing within structural Mountains of Indus Flysch and Ladakh Bathloliths along Indus Tsangpo Suture Zone (ITSZ) near Karu village (**d**) Thick deposits of lateral and end moraine positioned in a wide U-shaped valley carved by the Khardung glacier as evidence of past glacial activities near Ganglas village upstream of Leh towards Khardunga La Pass. (**e**) Huge deposits formed by the Aeolian processes on the mountain near Leh airport (**f**) Sand dunes deposits near Hunder village in Nubra valley, (**g**) Thick terrace of a paleolake deposit near Spituk Gompa. Flood plain and bar deposits of Indus River. (**h**) Huge Alluvial Fan emerging out from the Karakoram Mountains along the Shyok River Valley

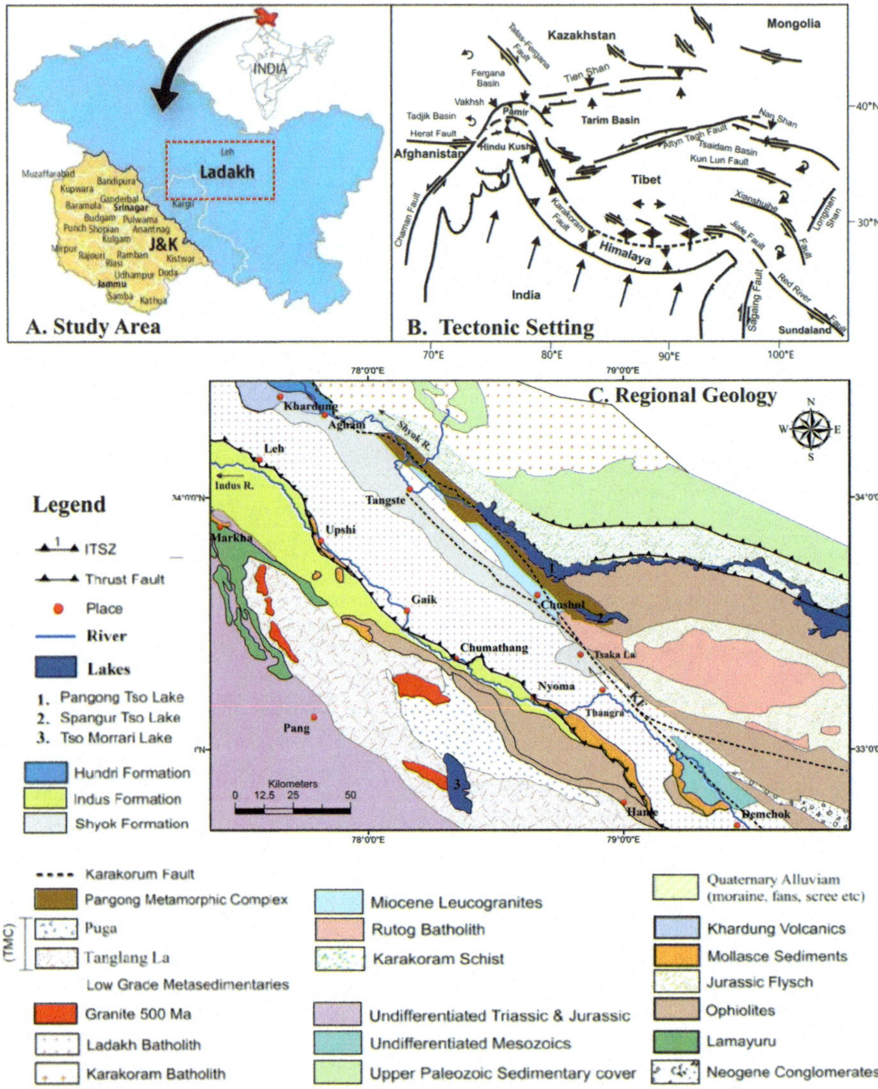

Fig. 2 Regional Geological cum tectonic map of northern margin of NW Himalaya, Ladakh Himalaya and Karakoram Zone: Inset (**a**) Division of Jammu Kashmir state into two Union territory showing extent of Ladakh encompassing the study area highlighted by red rectangle, (**b**) Tectonic framework of the Himalaya and Trans-Himalayan region depicting regional tectonic plates and fault lines, (**c**) Geological Map of Ladakh and adjoining areas, detailing geological formations with a color-coded legend and key features like Karakoram Fault, rivers and lakes such as Pangong Tso. (Lal et al. 2019; compiled from Thakur and Misra 1984; Thakur and Rawat 1992; Jain and Singh 2009; Thanh and Sajeev 2011; Dunlap and Wysoczanski 2002; Searle 2009)

revealing the hidden secrets of the India-Eurasia collision. These zones actually preserve evidences of Oceanic plate; emplacement of magmatic plutonic rock complex and final closure of Tethyan ocean (ISZ/ITSZ) (Honegger et al. 1982; Hodges 2000; Jain and Singh 2009).

The entire region of Ladakh is in the rain shadow of the NW Himalaya and has an arid to hyper-arid climate (Bookhagen et al. 2005; Burbank and Bookhagen 2006) and lies in the boundary zone between two major climatic circulations, the mid-latitude westerlies coming from the north and the Indian summer monsoon from the south (Benn and Owen 1998; Blöthe et al. 2014). The region shows wide diurnal and seasonal fluctuation between +40 ° to −40 ° giving rise to condition of both hostile cold climate like Polar Regions and aridity like deserts. Therefore, Ladakh is often called as *Cold Desert* having climate which is deficient in oxygen (Juyal 2014). The most accessible time to visit the area is from July to September as extreme cold climatic conditions prevailing rest of the year. Winter westerly winds, associated with the subtropical branch of the Northern Hemisphere westerly jet bring moisture from the Mediterranean region, which is the dominant source of precipitation in the region in the form of snow. Leh (capital of Ladakh) is a rain deficient area, and the average annual rainfall is about 10 cm/year. Scientific studies show that south Asian summer monsoon supplies between 250 and 500 mm of precipitation annually (Bookhagen et al. 2005). Enhanced monsoon influence during times of increased insolation in the past likely played a dominant role in controlling lake levels for many lakes throughout Tibet (Shi et al. 2001). The large proportion of rainwater is carried down the valleys during the short period of rainy season, and small part of rainfall percolates down through the soil and rocks to recharge mountain springs.

Thus, this area provide information about the dynamic relationship between climate and tectonics along the Indian-Eurasian collision boundary. This region is highly sensitive to climate change and its geological archives can provide reliable inputs on the causative factors of past climate events (Owen and Caffee 2006; Burbank and Bookhagen 2006); Kumar & Srivastava (2017). Well-developed lateral moraines preserved at many places in the region (viz. Bara Lacha La and Shyok and Nubra valleys) (Fig. 1), clearly indicate that this region experienced at least two glacial advances during the Quaternary period. Scientific studies show that the Tso Pangong (one of the salt water lakes in the area in the SE Ladakh) is in a closed basin, but it fed the Tangtse River during the Pleistocene time. The Tangtse River now receives most of its waters from melting glaciers on the northern side of the Pangong Range and waters from the Loi-Yogma valley (Shi et al. 2001; Pant et al. 2005). Besides Pangong Tso, there are numerous saline and fresh water lakes as well as many paleolake deposits which, besides the spectacular landscapes, also preserve interesting geoscientific stories and therefore form the subject matter of this article.

2 Regional Geological and Tectonic Setting

Geologically, the Trans Himalayan zone can be subdivided into four units which can be seen in the Ladakh region situated in the NW Trans Himalayas and bordering in the south with the Higher Himalayas (Thakur and Rawat 1992; Jain and Singh 2008; Searle et al. 1988; Searle 2015; Bohon et al. 2018). The major geological units present in the Trans-Himalaya in the Ladakh Region include Ladakh granite of the Island arc affinity, the ophiolites and subordinate metapelites (Fig. 2c). Rocks of the Trans-Himalaya in Ladakh region has been divided into five stratigraphic units from Zanskar in the south to Karakoram in the north listed below (Jain et al. 2005; Jain and Singh 2008; Searle and Tirrul 1991; Searle 2015; Thanh et al. 2011).

1.	Tso Moriri Crystallines (TMC):	Tanglang La formation
		Puga gneiss (with eclogites as enclosed boudins)
2.	Indus Tsangpo suture zone (ITSZ):	Kargil formation
		Indus group
		Nidar ophiolite
		Zildat Ophiolitic mélange
		Dras volcanic
		Lamayuru formation
3.	Ladakh Himalayas:	Ladakh batholiths
		Khardung Volcanics
4.	Shyok group:	Kole Molasse
		Diong formation
		Shyok ophiolite
5.	Karakoram units:	Low grade Tethyan sedimentary rocks
		Karakoram plutonic complex
		Karakoram metamorphic complex:
		(Tangtse group and Pangong group)

One of the remarkable tectonic elements present is a ~ 800 km long dextral strike-slip fault, known as Karakoram fault (KF). Geologically, the KF fault located within two sutures (SSZ and ITSZ) and cuts obliquely across the geologic terranes of the Pamir, Karakoram–Qiangtang, and Ladakh—Gangdese Ranges. In SE Ladakh it forms one major strand along the Nubra Valley and then bifurcates into the Pangong strand (considered to be geologically active) to the north east and the Tangtse strand in the south (Phillips 2008; Ravikant 2006). Between the two strands, the Pangong range is an exhumed metamorphic complex (Fig. 2). The Tangtse fault divides the pre-collisional Ladakh granitoids and Khardung calc-alkaline volcanic rocks to the SW from the metamorphic and magmatic rocks of the Karakoram to the NE and the Pangong fault cuts solely through rocks of the Karakoram terrane.

The offset/movement along the Karakoram fault is reported to vary from~40–150 km with slip rates varying between 4 and 40 mm/yr. (Avouac and Tapponier 1993; Searle et al. 1998; Brown et al. 2002; Phillips et al. 2004; Searle and Phillips 2007; Wright et al. 2004; Jade et al. 2004; Chevalier et al. 2005; Banerjee and Burgmann 2002; Phillips and Searle 2007). Prominent linear valleys, clearly recognizable on satellite imagery, follow the Karakoram fault trace (e.g., Liu 1993; Molnar and Tapponnier 1978). Fresh scarps mark offsets of alluvial and debris-flow can easily be seen on the ground and from high-resolution satellite imagery (e.g., Brown et al. 2002). Other than the hard rocks the area comprises thick alluvium deposits formed during Quaternary time which occur as an isolated and incomplete sedimentary sequence of river, paleo-lacustrine and glacial deposits.

3 Geomorphological Setup

The landscape of Ladakh is a reflection of the interaction among climate, geology and structures. Geomorphological architecture of this region is marked by the occurrence of diverse types of deposits in the area from extreme cold (moraines) to warm (fluvial) conditions and calms (Lake) to high energy (alluvial fans). The records of such varied environments signifies the influence of climate and tectonics on the landscape evolution of Ladakh. The Quaternary sedimentary deposits are scattered in the valley at different levels offering scope to infer the paleoclimatic changes and understand surface processes.

Indus is one of the oldest and longest rivers in the world, which originates from the northern slopes of the Mount Kailas at 5182 m amsl near lake Mansarovar in Tibet draining several rock formations (Clift et al. 2001). Indus is mainly nourished by glacier fed and snow fed tributaries such as Shyok, Nubra, Tangste in the north and Hanle, Zanskar in the south besides rainfall and runoff from valley glaciers.

Initially for ~210 kms, the Indus River follows along the NW-SE trend of the Karakoram Fault (Fig. 2c) and thereafter, it takes a westward 90° turn near Thangra village and flows through the high elevated plateau area of Ladakh and joins the Indus Tsangpo Suture Zone (ITSZ) with localized departures from this trend (Shroder and Bishop 1999; Searle and Owen 1999). The longitudinal profile of the Indus River along the Karakorum fault (KF) between Demchok and Thangra village and along the Indus Tsangpo Suture Zone (ITSZ) between Thangra till Mahe village shows gentle slope and at places water appears to be stagnant. Further downstream, valley widens and river shows entrenched to meandering nature and between Phey and Nimmo it flows into narrow bedrock valley in with sharp meanders after abandoning its course along the ITSZ.

Entire area exhibits geomorphologic units in a younging sequence comprising Structural hills>Intermontane valley>Alluvial fans (huge bajada surface), Glacial landforms and deposits (thick moraines, wide glaciated valleys) > Terraces (Fluvial and Paleolake) > Aeolian surfaces and deposits (sand dune) > Active river channel and its flood plains (Norin 1925; Pant et al. 2005; Owen and Caffee 2006; Phartiyal et al. 2005, 2013; Sant et al. 2011; Nag et al. 2015, Nag and Phartiyal 2015; Mujtaba et al. 2017; Kumar et al. 2017; Lal et al. 2019; Moga et al. 2019; Shukla et al. 2020).

Structural hills of a Ladakh batholith, Zanskar meta-sedimentaries, rocks of Indus formation, Ophiolitic suite and multiple volcanic episodes forms the oldest morphostratigraphic unit of the area ranging between ~6648 m and 3000 m amsl (Fig. 1c). Presence of thick moraines preserved at various places in the region (viz. Leh-Ganglas, Tangste, Pangong, Shyok-Nubra valleys) marks ice extent and serves as prime evidence of past glacial episodes (Fig. 1d). Available chronological records clearly indicate that this region has experienced multiple glacial advances during the Quaternary period (Owen et al. 2002; Owen and Caffee 2006; Dortch et al. 2010, 2013; Shukla et al. 2020).

The aridity of the region similar to desert conditions leads to the accumulation of thick pile of Aeolian sediments covering the hill slopes forming large sand ramps as seen near Spituk (Fig. 1e) and several other localities like Shey and Nyoma (Kumar et al. 2017). Numerous Crescent and Barchanoid shape sand dunes with wind ripple marks formed within river floodplain, abandoned fan surfaces and wide intermountainvalley floors (Fig. 1f).

Several terraces of glacio-fluvial, fluvio-lacustrine and lacustrine nature are well documented, majority of which are preserved in a linear stretch all along the Indus River between Nyoma and Saspol and other tributary river valleys like Zanskar, Pangong, Tangste and Shyok etc. These terraces are dominantly made up of medium-coarse sand, gritty sand, pebble beds, interfingering of local debris material and thick clay beds which are buried under debris at places.

Lacustrine clays are common as archives of paleolake deposits and are reliable indicator of causative processes, past climate and tectonics of the area (Searle 1996; Qayyum et al. 1997; Friend 1998; Clift et al. 2000). Lithologically, the paleolake deposits are characterized by laminated and bedded sediments that mainly comprise clay and silt sized particles deposited in very calm environment resulting in local development of varves. One such succession of sediments is exposed on the right bank of Indus between ~3210 and 3280 m amsl near Spituk Gompa (Fig. 1g). Evidences of these paleolake deposits occur as either remnant patches attached on the sides of river valley or isolated terraces and as significant valley fillings/depositional sequences which mark localized damming of the streams and low energy conditions suitable for deposition of lacustrine sediments (Fig. 3).

Occurrences of prominent alluvial fans forming even and highland surfaces can be seen deposited in front of streams emerging from tributary valleys and abundant

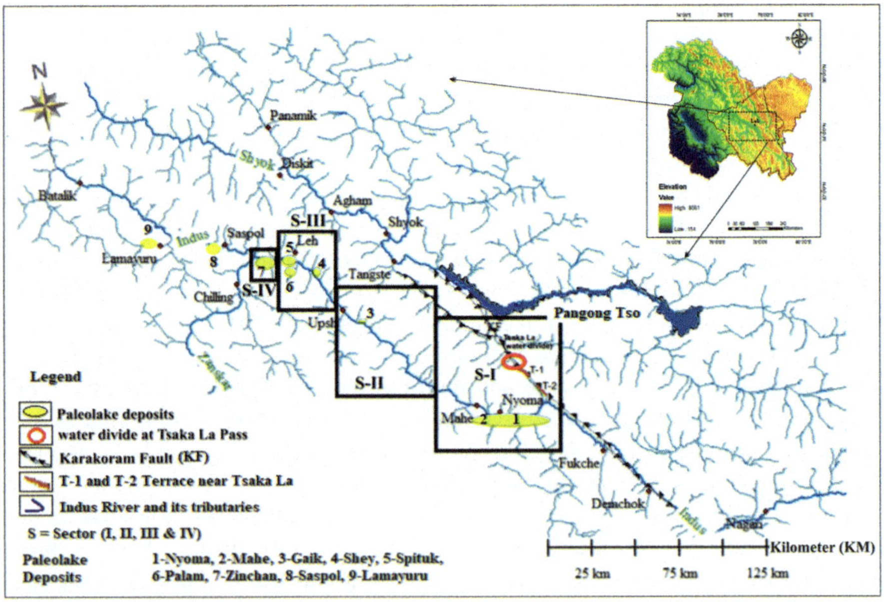

Fig. 3 Drainage map of the Ladakh showing distribution of paleolake deposits at various altitude level situated along the course of Indus River and its tributaries, water divide at Tsaka La Pass and traverse of Karakorum fault (modified from Lal et al. 2019)

colluvial fans as gravity-controlled deposits along the Indus River and its tributaries (Fig. 1h). One of the well-developed landforms is the Bajada surface formed by coalescing fans and it continues for around 30 km in form of 0.5–2 km wide areas with vertical cliff face (~50 m thick) along the left bank of Indus River between Karu and Phey. Two sources of sediments for the fans, Ladakh batholith and the Indus molasse were identified by Sangode et al. (2013).

4 Terrestrial Water Bodies of Past and Present

Ladakh region (along with whole NW Himalayas) host several fresh water as well as salt water lakes that serves as spectacular tourist sites across the globe such as Pangong Tso Lake close to Indo-China Border. Tso is the term regionally used for a lake. There are number of lakes present in Ladakh at different altitude levels with various dimensions, varied nature of water (fresh/saline) (Fig. 4). At places, these lakes also preserve relics of massive deposits of past lakes along Indus River and

Fig. 4 Glimpse of present lakes in Ladakh well-known tourist hotspot. (**a**) Landscape view of one of the most beautiful lakes in Ladakh "Pangong Tso" near Maan village close to the Indo-China Border. (**b**) Yarab Tso (lake), a small lake surrounded by rocky hills in Nubra Valley located near the Simur village. (**c**) A view of the Tso Moriri Lake from the Karzok village. (**d**) A panoramic view of the Lake Kiagar, a small high altitude remnant lake in the Rupshu Valley (wild Yak grazing near the lake can also be seen)

adjoining tributaries. These water bodies are mainly nourished by the glaciers and snow fed streams. In general, these lakes are formed as a result of reactivation of fault/s that led to the production of debris avalanches resulting from slope failures, thus, blocking the past drainages and ultimately forming the lakes (Gopal et al. 2002; Phartiyal et al. 2005, 2013 and reference therein).

Modern/Present Lakes in Ladakh

As mentioned earlier, one of the most prominent geological feature of this region is the presence of many Lakes including: Pangong Tso (33°55′17.26"N; 78°29′3.91″E); Tso Moriri (32°54′39.42"N;78°18′57.39″E); Tso Kar (33°18′8.10"N;78°1′3.75″E); Yarab Tso (34°44′50.05"N;77°33′8.05″E); Mirpal Tso (33°27′47.82"N;78°36′44.09″E) and Kiagar Tso (33°5′55.54"N;78°17′58.65″E). Among all these lakes Pangong Tso and Tso Moriri are the two biggest and famous lake sites and are described in more details here.

Pangong Tso is the largest lake in the area with a catchment area of 2000 km^2. It is ~134 km long and extends from India to China at the height of 4250 m above the sea level with its aerial extent being around 650 sq. km; the maximum width of the lake being ~8 km on the Indian side and an average width of ~3 km and ~ 40-m depth (Fig. 4). Geographically, it is not a part of the Indus River basin area and is a separate land locked river basin without any outlet, making it saline. The lake receives water from snow-fed streams and has moderate alkaline water of pH~ >9. Depending on the length of the melting season, fresh water input alternating between saline/brackish and fresh water, modulates the lake level and its pH conditions (Srivastava et al. 2020). Based on former shorelines of the lake multiple paleolake levels have been marked which serves as a strong evidence for showing regional climatic fluctuation reported to be ~4280 m to 4320 m amsl and have been dated to be ~35 and 45 ka (Norin 1925; Huang et al. 1989; Shi et al. 2001; Pant et al. 2005).

The formation of the Pangong Tso basin has been attributed to the damming of an old outlet by fans from tributary torrents (Rathour et al. 2020; Majeed et al. 2021). The lake receives meltwater of the glacier in summers, while the loss of water occurs only through evaporation as it is landlocked lake. Though, earlier on, Pangong Tso had an outlet to the Shyok River, a tributary of the Indus from the northwest, now closed due to natural damming.

The north-western basin of Pangong Tso is structurally controlled by oblique slip faulting along the Pangong strand of the Karakoram Fault (Dunlap et al. 1998; Searle et al. 1998; Searle and Phillips 2007). Scientific study by previous workers also revealed that this lake occupies a long, submerged valley which has been dammed to the west by a ridge formed by tectonic activity associated with the Karakoram strike slip faulting or by a moraine deposit during Last Glacial Maximum (Norin 1925; Huang et al. 1989; Phartiyal et al. 2005). It has been studied that during Pleistocene the lake used to feed Tangtse river and that the partial drainage of Pangong Tso resulted in a catastrophic flood. The flood was possibly due to seismic activity along the Karakoram Fault in combination with an intensified monsoon phase at 9.6–11 ka leading to an abrupt opening of the hydrological system at 11 ka (Dortch et al. 2011).

Tso Moriri is situated at an elevation of about 4900 m south of the Indus River, which originates near the Mansarovar Tso in Tibet. The melt water of ice masses left behind by the retreating glaciers has probably formed the Tso. It is one of the largest lakes in Ladakh region and like Pangong Tso is almost like an inland sea (Fig. 4). Geographically, the Tso Moriri area is situated in the eastern Ladakh; about 220 km SE of Leh; within latitude of 32°40′ and 35°25’ N and longitude of 77°50′ and 78°40′ E. The lake is situated near a village called Karzok, which is connected by a non-metallic road to a small town called Upshi on the bank of the Indus River. It has a length of about 25 km and a varying width of 6 to 8 km with a depth of more than 50 m at the deepest point (Shukla et al. 2012).

The lake is fed by a number of small glacial streams and has no external drainage. This fact coupled with a high rate of evaporation, has resulted in the waters being brackish and bereft of any life. Scientific study revealed that this lake is formed in the Late Quaternary due to the damming of the NNW–SSE trending River and its tributaries by debris avalanches/deposits initiated mainly by tilting of fan due to tectonic activity along the N–S to NNW–SSE trending fault passing up to Kaurik, Leo and Sangla valley (Phartiyal et al. 2005, 2013).

Lake Tsokar: The lake is situated south of the Indus Suture Zone in the northwestern Himalayas, approximately 160 km southeast of Leh and 80 km northwest of Tso Moriri by road. It is entirely brackish and covers an area of around 20 km², with a highly irregular shape. Positioned at an elevation of 4,530 meters (14,860 feet) above sea level, the lake receives water from several small streams and meltwater from the surrounding mountains. With no significant outflow, it is considered an endorheic (closed) lake (Phillip and Mazari 2000).

Lakes of Past (Paleolake Deposits)

A spread of paleolakes terraces and deposits have been reported along the Indus valley flanks in the Ladakh (Fig. 3). They are in general; lies at an altitude of >3000 m amsl. Some of the common sites of paleolakes in the area are around Nyoma, Mahe, Shey, Leh, Spituk, Zinchan, Saspol, lamayuru and Khalsar etc. A prominent valley fill occurs near Lamayuru and Nyoma and are shown in Fig. 5.

The existence and significant depositional sequences suggests that the lakes in this region may develop due to localized damming of the river courses either by climatic and tectonic induced landslides, aggrading alluvial fans and moraines left by advancing glaciers and low energy conditions suitable for deposition of lacustrine sediments. The chronological studies of these paleolake deposits have been carried out by several workers suggesting their Late Quaternary affinity (Burgisser and Gansser 1982; Fort et al. 1989; Sangode and Bagati 1995; Kotlia et al. 1997; Pant et al. 2005; Phartiyal et al. 2005, 2013; Phartiyal and Sharma 2009; Sangode et al. 2011, 2013; Blöthe et al. 2014; Nag et al. 2015; Mujtaba et al. 2017; Lal et al. 2019).

Some modern lakes also like Pangong Tso and Tso Moriri occupy the glacial-fluvial valleys in which water is impounded by detrital blockage caused by tectonic and or climatic disturbances. Some of these lakes have terraces and water marks indicating past fluctuations (Fontes et al. 1996, Gasse ct al. 1996).

Majority of these paleolakes are located near the collision zone (ITSZ and SSZ) and along regional and local thrusts/faults, has huge potential in terms of geological

Fig. 5 Relics of magnificent lake of Past (Paleolake deposits) spread all over in Ladakh at different altitude levels along Indus and its tributaries. (**a**) Remains of paleo-lacustrine and associated sediments at Lamayuru "Moonland of Ladakh" situated on the Srinagar-Leh highway. (**b**) Thick paleolake terrace near Zinchan nalla and Indus confluence between Phey and Nimo. (**c**) Vertical view of Spituk paleolake terrace showcasing laminated clays inter-bedded with alluvial strata, massive beds of compact yellow and varved clay on the right bank of River Indus near Spituk Gompa. (**d**) Massive yellow clay and varved clay beds of Spituk paleolake deposit on the right bank of river Indus near Spituk Gompa. (**e**) Field view showing spread of the lacustrine clay of Shey deposit on the right bank of Indus River heading the Shey Palace. (**f**) View of the Khalsar paleolake deposit from north pullu crossing Khardung La Pass. (**g**) Paleolake deposit near Mahe village placed along the valley walls on both the sides of the Indus River. (**h**) Around 6.5 m thick fluvio-lacustrine section exposed on the right bank of Indus River around 2.5 km SW of Nyoma village

heritage as remnants of these past magnificent lacustrine deposits hosts seismites (deformational structures that represent records of critical long term information of the past seismic activities during the Quaternary Period). The epicenter, frequency and magnitude of numerous past earthquakes occurrences can be better visualized and understood with the help of these peculiar indicators (Upadhyay 2001; Phartiyal et al. 2005; Mujtaba et al. 2017; Üner et al. 2017, 2019).

Formation of such lakes appears to have an implicit relation with the fluctuation in glaciations during Late Quaternary and thus the timings of glaciations–deglaciations assume significance in understanding the regional and global climate phenomenon (Owen et al. 1998, 2002; Finkel et al. 2003; Blöthe et al. 2014).

Small climate or tectonic induced activity can trigger a catastrophic event that can mobilise huge amount of loose and unconsolidated sediment material piled everywhere. Recent fluctuations reflected in the irregular and extreme precipitation within a short time interval at various places over the arid landscape of Ladakh has witnessed much devastation (2006, 2010, 2012 and 2015 flash-floods and cloudbursts) severely impacting the local people lives and enormous infrastructural damages (Hobley et al. 2012; Sangode et al. 2017; Suri 2018; Schmidt et al. 2020).

Recent studies highlighting importance of paleoflood deposits and glacial lake outburst floods to reconstruct past floods records in the Trans-Himalayan region of Ladakh have further shown the significance of these deposits (Ives 1986; Sangode et al. 2011; Ziegler et al. 2016; Majeed et al. 2021). The chronological data of these paleoflood deposits can reveals important of extreme climate shift corresponding to global climate fluctuations. Hence, evaluating the unique information stored within the present lakes and past lake deposits w.r.t. catastrophic and hazardous impact on the environment, landscape and society (Ives 1986; Coxon et al. 1996; Reynolds 1998; Richardson and Reynolds 2000; Dortch et al. 2011; Schmidt et al. 2020).

Some of the important paleolake sites have been described in details in the following section.

Lamayuru Paleolake "Moonland of Ladakh" (34°17′05" N; 76°47′00"E)
A prominent valley fill near Lamayuru at an elevation of ~3600 m in Zanskar range of the Ladakh region, represents significant ~200 m thick paleolake deposit comprising lacustrine clay and associated sediments located (Fig. 5a). The high clay content of the bedrock and cold arid climate played a key role in developing a peculiar badland landscape, called Moonland, within the area of northern half of the lake floor sediments (Fort et al. 1989; Shukla et al. 2002; Juyal 2014; Moga et al. 2019).

The Lamayuru palaeolake deposit were lithologically described and dated to be late Quaternary with different dating methods (Bagati and Thakur 1993; Kotlia et al. 1997, 1998; Blöthe et al. 2014; Nag and Phartiyal 2015). The results indicate that Lake Lamayuru was formed during a cold and humid climate, which according to dating of the lacustrine sediments have persisted until 25,000 years BP (Nag and Phartiyal 2015).

The Spituk Paleolake (34°07′ 50"N; 77°31′ 30" E)
A ~ 65-meter-thick succession of interlayered fluvio-glacial-lacustrine sediments is exposed near Spituk Gompa on the right bank of the Indus River between ~3210 and 3280 m amsl. It is one of the most well developed and well preserved Quaternary

sedimentary record in this sector. Based upon the gross sediment association, the succession can be divided in three units (U1, U2and U3). The lower unit (U1) is ~25 m thick and is divisible into a lower fine sandy zone, a middle silt-clayey zone and an upper coarse sandy zone. The lower sandy zone has layers of horizontally and ripple-laminated channel sand, flame structure and convolute bedding formed in a channel environment.

Middle Unit (U2), consist of 20 m thick lacustrine clays exposed in the road cutting. In the lower part (~5 m thick) of unit 2, clay is massive and hard with few centimeters thick pinkish bands at around 50 cm intervals. In upper part (~15 m thick), clay is thinly bedded and finely laminated with a 30–90 cm thick gritty sand layer towards the top (Fig. 5d). The gritty layer has a sharp and wavy lower contact. The clay above the gritty bed has semi-vertical joints. Top surface of the clay unit has rounded nodular concretionary structures.

Top Unit (U3) is dominantly fluvial in origin and consists of gritty sand, medium to coarse sand and silt-clay. It is ~18 m thick and has a mound like appearance due to subsequent wind action. (Fig. 5c).

Shey Paleolake Section (34°04′29.2“ N; 77°38’ 08.0” E)

A paleolake deposit is present near Shey Palace on the right bank of the Indus river (Fig. 5e) near Shey monastery. It is about 15 km upstream of the Spituk paleolake deposit. The deposit comprises around ~30-meter-thick sequence of fluvio-glacial-lacustrine beds. The base of deposit lies at an elevation of 3252 m amsl about 12 m above the modern base level of Indus.

The lithofacies variation in this section expresses dynamic change in the depositional environment regime which is possibly due to climatic oscillation. Shey paleolake deposits mainly comprise clay and silt sized particles deposited in very calm environment.

Shey deposit consists of interfingering of clay and fine sand/silt at the base followed by thick clay deposits (laminated and massive) and fluvial sand, indicated calm environmental condition. This is overlain by fan deposits which are covered by aeolian deposits at the top. Lower section of the lower part has some biological activity while upper section is oxidized. Pen-Contemporaneous Deformation (PCD) is observed in form of sand dome, clay ball pseudonodules, tilting of beds and fault and undulation in the middle clay unit of the section (Fig. 6d). Overall lithofacies variation in this section expresses change in the depositional environment regime which is possibly due to climatic oscillation. Chronological evidences of this lake deposits were dating to be ~91 ka (Lal et al. 2019).

Zinchan Paleolake Deposit (34°07′ 11.9“N, 77°25’ 16.9”E)

A ~ 50 m thick north facing fluvio-lacustrine sedimentary sequence preserved in the Nimmo gorge on the left bank of Indus River where the Zinchan nalla join the Indus River before the confluence with Zanskar river at Nimmo. Entire deposits lie over the bedrock gorge of Indus River. Lower parts of this section are very difficult to access on being a cliff face on a steep slope therefore ~9 m thick section exposed along the road was studied. The top part of lake deposit is defined by a 3260 m contour level. Lower part is fluvial dominant and lacustrine while the middle part shows structures indicating neotectonic activity.

Fig. 6 Geomorphic evidence of tectonic induced sedimentary structures formed within Late Quaternary deposits along the Indus Tsangpo suture zone (ITSZ) and the Karakorum fault (KF) (**a**) Local thrusting observed in clay beds of paleolake deposits near Durbuk village (Brown et al. 2002) (**b**) Fault scarp in an alluvial fan indicating active faulting near Panamik on the eastern side of Nubra valley (**c**) Spectacular Box fold developed during seismically triggered soft sediment deformation (SSD) within Plaeolake deposits in the Tangste valley. (**d**) Folding within the laminated clay bed overlain by a horizontal clay bed. (**e**) Some beds show intrusion of clay dykes and at some places clay is found to be intruding within the overlying sediments as clay dykes and sills. (**f**) Faults within the Zinchan sedimentary deposits are convincing evidence of neo-tectonic activities during Quaternary. Note: Fig. 6 D, E and F shows evidence of syn-post depositional deformations within Zinchan Paleolake deposits between Phey and Nimmu along ITSZ (Mujtaba et al. 2017)

In gorge area, these deposits preserve strongest evidence of syn-post depositional deformation structures characterized by deformation, scouring, cross cutting and liquefaction features. The imprints of tectonism are represented by several minor thrust/fault developed perpendicular to ITSZ and their imprints are well preserved within the deposits (Fig. 6f). All along the road, imprints of neotectonic activities are observed. Clay and sand dykes are also noted (Fig. 6g). Clay beds show excellent folding and slump structures (Fig. 6e). The beds are usually disturbed, displaced and deformed showing, folding, faulting and tilting of the clay beds indicating tectonic instability. No evidence of the damming of the river is preserved in the valley. From the middle and top of bedded silt-clay along road section near Zinchan nalla confluence ages of 125 ± 11 and 87 ± 8 ka have been obtained (Mujtaba et al. 2017).

Khalsar Paleolake Deposit (34°29′32.83"N, 77°42′7.18″E)
About 150 m thick fluvio-lacustrine sedimentary deposit is exposed near Kalsar village crossing Khardung La Pass along the left bank of the Shyok river valley (Fig. 5f). The overall sedimentological composition of this deposit includes flat-lying, thinly bedded clays, silts and fine- to coarse-grained sands, conglomerates and breccias (Upadhyay 2001) which were found similar to a 20 m thick exposed section along the right bank of the Shyok river near the Tirit village. Beside compositional Soft-sediment deformation structures (SSD) have also been reported from this deposit that occurs in several well-defined intervals and are observed laterally continuous for hundreds of meters (Upadhyay 2001, 2003; Phartiyal et al. 2005; Phartiyal and Sharma 2009).

Mahe Paleolake Deposit (33°15′50.30"N; 78°30′39.99″E)
The Indus River makes a prominent loop near Mahe in which a ~ 6 m thick lacustrine sequence is developed on the right side of river (Fig. 5g). In the lower part, it comprises off-white clays showing crude horizontal laminations. It is overlain by khaki clays with horizontal fabric. The contact is definable but not sharp. The khakhi clays pass upward in khakhi-brown clays characterized by vertical fabric. The Optically Stimulated Luminescence (OSL) age recorded 41 ± 2ka and 14 ± 2 ka for the samples from this area (Kumar et al. 2017).

Nyoma Paleolake Section (33°12′ 06.3" N; 78°37' 28.5" E)
At Nyoma~6.5 m thick fluvio-lacustrine section is located at an elevation of 4157 m amsl on the right bank of Indus River around 2.5 km SW of Nyoma village. Base of the section consists of alluvial fan deposit comprising of poorly sorted, angular and disorganized clasts in sandy-gritty matrix. Lower fan unit is overlain by fluvial facies followed by lacustrine unit (clay-silt in the middle part) showing interfingering of fluvial and lacustrine facies (Fig. 5h).

The basal fan unit of the section is ~0.25 m thick and consists of coarse grained, sub angular, light purple clasts (which may be the clasts of meta-sediments of flysch deposit). These deposits are poorly sorted and without imbrications and represent fan cut terrace. Middle part of the section, ~3 m thick unit shows clay silt interfingering representing lacustrine deposit. It has a ~ 1 m thick laminated clay (yellow with pink tinge) with 0.5 m thick grey fine silty sand unit overlying it. Top part of the middle unit comprises ~1.5 m thick silty clay unit with gritty sand above it. Within silty clay and gritty sand unit a wavy layer occurs in patches (Fig. 5h). Entire section is overlain by the fan debris similar to that was observed at the base and at places thinly laminated clay with sand dyke intrusion without disturbing overlying and underlying layer was noticed. One of the samples (NY-2) gave an age of 36 ± 3 ka.

5 Geological Significance of Lakes

Climate proxy parameters preserved in form of lake deposits provide important information about the climatic conditions as well as evolution of the geomorphic landscape of the Ladakh region.

5.1 Catastrophic Events Due to Karakoram Reactivation during Quaternary

Pangong Tso is the largest saline lake in the Ladakh region which has a strong geological significance as one of the strand of Karakoram fault passes through this area. Field studies provide evidence of an extreme catastrophic flood (dated ~9.6–11.0 ka) in the Tangtse valley caused due to overtopping through partial drainage of ~18.3 km3 of water which was suddenly released leading to modification of entire landscape. (Dortch et al. 2011; Phartiyal et al. 2015). The seismic reactivation of the Karakoram Fault along with an intensified monsoon could be a probable cause for such abrupt event.

Other geomorphic evidence such as local thrusting observed in clay beds of paleolake deposits near Durbuk village (Fig. 6a), fault scarp in an alluvial fan near Panamik on the eastern side of the Nubra valley (Fig. 6b) and box fold developed during seismically triggered soft sediment deformation (SSD) within Plaeolake deposits in Tangste valley (Fig. 6c) preserves imprints of neo-tectonism along the Karakorum fault (Brown et al. 2002; Imsong et al. 2017).

5.2 Switching of Lake Sediment Provenance and Drainage Disorganization

Switching of sediment provenance and drainage disorganization is another important aspect these lakes which was inferred from the mineralogical studies of paleolake deposits along the Indus river (Spituk and Shey) which revealed three major suites of rocks with distinctive heavy mineral assemblages. First, the granite and granite gneiss of the Ladakh Group is exposed in the largest part of the basin and can contribute biotite as the major and amphiboles, zircon/monazite and magnetite as sub-ordinate heavy minerals to the Quaternary deposits (Sharma 1992). The second suite consisting of ophiolites and other mafic to ultramafic rocks are presented in both the northern and the southern suture zones (ITSZ and SSZ). Third, the metapelitic rocks composed primarily of amphibolite facies metapelite minerals such as staurolite, kyanite and garnet. A possible source of metapelitic minerals could be the Tso Moriri Crystalline Complex near lake Tso Moriri drained by a tributary that joins the Indus River near Mahe village (Munack et al. 2014; Guillot et al. 1997, 2000; Sachan et al. 2004) and the Karakoram Metamorphic Complex near Pangong Tso lake (Thanh et al. 2011) which is not connected to the Indus River course at present.

Metapelitic minerals (typically kyanite and/or garnet and/or staurolite) occur throughout the Shey deposit while these are present only in the lower unit of the Spituk sequence. Therefore, mineralogically, the lower Spituk and the Shey deposit suggest similarity of provenance. Firstly, this hypothesis stands verified if the age of the Shey sequence is equivalent or older to the lower part of the Spituk sequence (lower unit ~70 ka). This is borne out by the age of Shey sequence which is ~90 ka (Lal et al. 2019). Secondly, the amphibolite facies metapelite minerals such as staurolite, kyanite and garnet are present in lower Unit1 while are significantly absent in middle Unit2 and upper Unit3 of the Spituk deposit. These assemblages indicate a change of sediment source from magmatic and metamorphic rocks to predominantly magmatic rocks. Nearest outcrops of metapelites lie in the Tangtse Valley (Pangong Metamorphic Complex) which does not presently drain in the Indus River.

As described by Lal et al. (2019), topographically, a valley connection exists between the two but a subdued water divide in the valley near Tsaka La has obstructed the flow of water from Tangste to Indus. They interpreted the drainage disruption due to the accumulation of moraine or dislocation of landforms due to the activity of Karakoram Fault (Fig. 7).

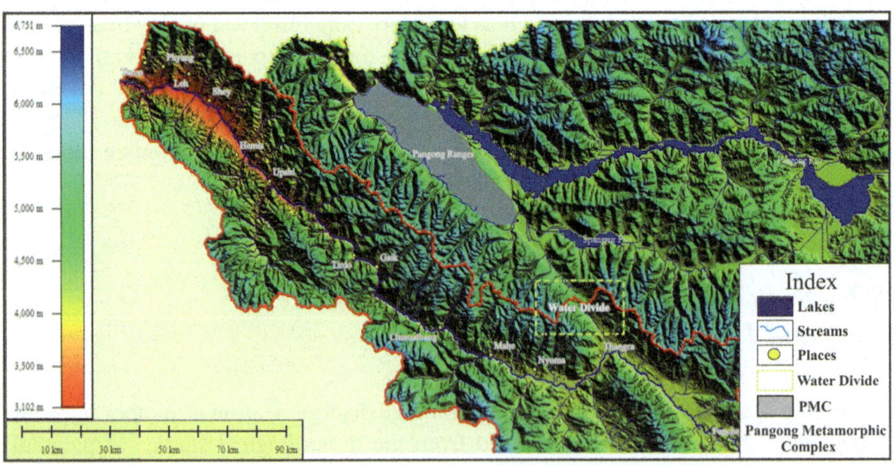

Fig. 7 SRTM derived digital elevation model (DEM) showing the broad view of the water divide region (within dotted yellow rectangle) where the PMC is marked in grey colour present near Pangong Tso lake. Cross sectional profile across Indus River in different sectors from upstream Nyoma to Zinchan in downstream showing wide variations in valley width and river behaviour

Considering this and other studies in conjunction with the documented slip-rates of the Karakoram fault it can be said that it may have played a vital role in drainage disruption. Similar tectonic activity induced damming involving the Karakoram fault has been envisaged for the Pangong Tso lake (Huang et al. 1989; Phartiyal et al. 2015).

5.3 *Palaeoclimatic Stratigraphy of Ladakh Region in India*

On account of the climate cum geographic position of the Ladakh it is worthwhile to evaluate the influence of global and regional climate forces. To facilitate this evaluation, the fluvial, galcial and lake deposits serves as important parameters to represent glacial (cold) and interglacial (warm) periods respectively during the Quaternary time. The glacial periods given by Owen et al. 2002 and new age data (Fig. 8) is plotted with respect to the global climate record expressed in terms of $\delta^{18}O$ as shown in MIS record (Lisiecki and Raymo 2005). Blue colour indicates glacial, red colour indicates interglacial period and white colour marks no record/ lacuna which still need further inputs for reconstructing high resolution past climatic terrestrial stratigraphy of Ladakh.

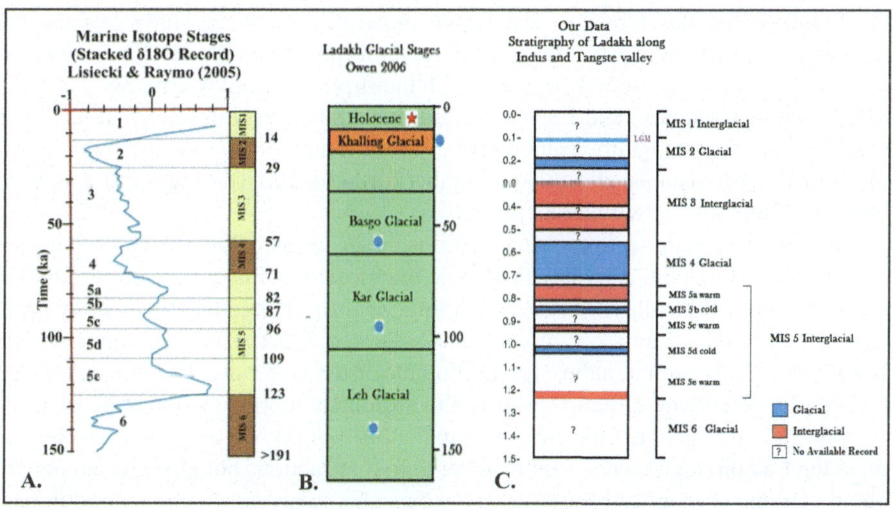

Fig. 8 Global climate record (MIS) vs. Late Quaternary stratigraphy of Ladakh based on the dating of sediments associated to glacial, fluvial and lacustrine deposits. Panel A shows a graph of marine isotope stages (MIS) over time, highlighting stages MIS 1 to MIS 6 (Lisiecki and Raymo 2005). Panel B illustrates Ladakh glacial stages, including the Holocene, Khalling, Basgo, Kar, and Leh glacial periods (Owen et al. 2002). Panel C presents Late Quaternary stratigraphy of Ladakh, indicating interglacial and glacial periods with color coding

6 Concluding Remarks

The present-day lakes as well as remnants of paleolakes not only represent a spectacular geological feature but also preserve remarkable geological stories of climate tectonic interactions in the uniquely located Ladakh plateau.

Variation in mineralogical and textural features along the depositional section at Spituk paleolake section having a lower unit comprising of ~fine sand+silty clay indicating channel environment; middle unit mainly comprises of laminated silt clay and massive clay bed indicative of cold lacustrine environment and the uppermost unit comprises of horizontally laminated gritty sand with lenses of grey coarse sand denoting a high-energy fluvial environment is an excellent example which is very well preserved close to the Leh airport. A lake phase of over 30 ka is exposed in this section.

Similar depositional environment inferences are also drawn from others sections e.g., Palam section and Phey section with OSL age ~ 74 ka and 79 ka respectively (Lal et al. 2019). Both of these sections are easily approachable and represent very good geotourism sites conserving significant geoheritage. Study of paleolake deposits suggests that there was warm fluvial environment during ~78-72 ka time (that actually corresponds with the beginning of sedimentation in the valley), followed by fully established lake environment at around ~57 ka, which was again followed by warm environment at 49-46 ka.

An even more important geoheritage is preserved in the heavy mineral records of the Spituk paleolake deposits. The heavy mineral provenance study reveals that there occurs a change in assemblage from metamorphic to magmatic. The metamorphic source must be from Karakoram Metamorphic Complex (mainly Pangong Metamorphic Complex) and Ladakh batholith as the magmatic source. It has been observed that the metapelitic source actively contributed till ~73 ka after which supply from this provenance abruptly ended that further is related to induced damming because of tectonic activity along the KF.

Imprint of climate-tectonic interrelationship in terms of lacustrine/fluvial processes effected by fault activity can also be manifested at Tsaka La section. The KF passes through the valley that is connecting with the Pangong Tso lake area and hosting the Tsaka La water divide. The reported Quaternary slip rates of the Karakoram Fault and the nature (glaciofluvial) of the water divide deposit at Tsaka La suggests a tectonic-climate control of the drainage reorganization at ~73 ka.

Thus, it is justified that the Geo-sites in the form of present and paleolake deposits in the Ladakh regions not only provide tourist attractions but also give an insight about extreme geological processes in the past and paleo-environment conditions.

Acknowledgements The authors are thankful to Prof Naresh Chandra Pant for inviting the paper and constructively suggesting the changes. Authors also thank Dr. H.S. Saini for help in field campaigns as well in discussions.

References

Avouac JP, Tapponier P (1993) Kinematic model of active deformation in Central Asia. Geophys Res Lett 20(10):895–898

Bagati TN, Thakur VC (1993) Quaternary basins of Ladakh and Lahul–Spitiin north-western Himalaya. Curr Sci 64:898–903

Banerjee P, Burgmann R (2002) Convergence across the northwest Himalaya from GPS measurements. Geophys Res Lett 29:13

Benn DI, Owen LA (1998) The role of the Indian summer monsoon and the mid-latitude westerlies in Himalayan glaciation: review and speculative discussion. J Geol Soc 155:353–363

Blöthe JH, Munack H, Korup O, Fülling A, Garzanti E, Resentini A, Kubik PW (2014) Late quaternary valley infill and dissection in the Indus River, western Tibetan plateau margin. Quat Sci Rev 94:102–119

Bohon W, Hodges KV, Tripathy-Lang A, Arrowsmith JR, Edwards C (2018) Structural relationship between the Karakoram and Longmu co fault systems, southwestern Tibetan plateau, revealed by ASTER remote sensing. Geosphere 14(4):1837–1850

Bookhagen B, Thiede RC, Strecker MR (2005) Late quaternary intensified monsoon phases control landscape evolution in the northwest Himalaya. Geology 33:149–152

Brown ET, Bendick R, Bourles DL, Gaur V, Molnar P (2002) Slip rates of the Karakorum fault, Ladakh, India, determined using cosmic ray exposure dating of debris flows and moraines. J Geophys Res 107:2192

Burbank DW, Bookhagen B (2006) Topography, relief, and TRMM-derived rainfall variations along the Himalaya. Geophy Res Lett, Wiley Online Library 33

Burgisser HM, Gansser A (1982) Late glacial lake sediments of the Indus valley area, northwestern Himalayas. Eclogae Geol Helv 75(1):51–63

Chevalier M-L, Ryerson FJ, Tapponnier P, Finkel RC, Van Der Woerd J, Haibing L, Qing L (2005) Response to comment on "slip-rate measurements on the Karakorum fault may imply secular variations in fault motion". Science 309:1326c

Clift PD, Shimizu N, Layne GD, Blusztajn JS, Gaedicke C, Schlüter HU, Clark MK, Amjad S (2001) Development of the Indus fan and its significance for the erosional history of the Western Himalaya and Karakoram. Geol Soc Am Bull 113(8):1039–1051

Clift P, Shimizu N, Layne G, Gaedicke C, Schluter HU, Clark M, Amjad S (2000) Fifty five million years of Tibetan evolution recorded in the Indus fan. EOS Trans Am Geophys Union 81:277–281

Coxon P, Owen LA, Mitchell WA (1996) A late quaternary catas- trophic flood in the Lahul Himalayas. J Quat Sci 11:495–510

Dewey JF, Burke KCA (1973) Tibetan, Variscan and Precambrian basement reactivation: products of a continental collision. J Geol 81(6):683–692

Dortch JM, Owen LA, Caffee MW (2010) Quaternary glaciation in the Nubra and Shyok valley confluence, northernmost Ladakh, India. Quat Res 74(1):132–144

Dortch JM, Owen LA, Caffee MW (2013) Timing and climatic drivers for glaciation across semi-arid western Himalayan–Tibetan orogen. Quat Sci Rev 78:188–208

Dortch JM, Owen LA, Schoenbohm LM, Caffee MW (2011) Asymmetrical erosion and morphological development of the Central Ladakh range, northern India. Geomorphology 135:167–180

Dunlap WJ, Weinberg RF, Searle MP (1998) Karakoram fault zone rocks cool in two phases. J Geol Soc Lond 155:903–912. https://doi.org/10.1144/gsjgs.155.6.0903

Dunlap WJ, Wysoczanski R (2002) Thermal evidence for early Cretaceous metamorphism in the Shyok suture zone and age of the Khardung volcanic rocks, Ladakh, India. Journal of Asian Earth Sciences 20(5):481–490

Finkel RC, Owen LA, Barnard PL, Caffee MW (2003) Beryllium-10 dating of Mount Everest moraines indicates a strong monsoon influence and glacial synchroneity throughout the Himalaya. Geology 31(6):561–564

Fontes JC, Gasse F, Gilbert E, Derbyshire E (1996) Holocene environmental changes in Lake Bangong Basin (western Tibet); part 1, chronology and stable isotopes of carbonates of a Holocene lacustrine core. Palaeogeogr Palaeoclimatol Palaeoecol 120:25–47

Fort M, Burbank DW, Freytet P (1989) Lacustrine sedimentation in a semi-arid alpine setting: an example from Ladakh, northwestern Himalaya. Quat Res 31:332–352

Friend PF (1998) General form and age of the denudation system of the Himalaya. GFF 120(2):231–236

Gasse F, Fontes JC, Van Campo E, Wei K (1996) Holocene environmental changes in Bangong Co basin (Western Tibet); part 4, discussion and conclusions. Palaeogeogr Palaeoclimatol Palaeoecol 120(1–2):79–92

Gopal B, Chauhan M, Zutshi DP (2002) High altitude lakes of Ladakh: changes since the Yale North India expedition. Internationale Vereinigungfürtheoretische und angewandte Limnologie: Verhandlungen 28(2):519–523

Guillot S, De Sigoyer J, Lardeaux JM, Mascle G (1997) Eclogitic metasediments from the Tso Morari area (Ladakh, Himalaya): evidence for continental subduction during India-Asia convergence. Contrib Mineral Petrol 128(2):197–212

Guillot S, Hattori KH, de Sigoyer J (2000) Mantle wedge serpentinization and exhumation of eclogites: insights from eastern Ladakh, northwest Himalaya. Geology 28(3):199–202

Hobley DE, Sinclair HD, Mudd SM (2012) Reconstruction of a major storm event from its geomorphic signature: the Ladakh floods. Geology 40(6):483–486

Hodges KV (2000) Tectonics of the Himalayas and southern Tibet from two perspectives. Geol Soc Am Bull 112(3):324–350

Honegger K, Dietrich V, Frank W, Gansser A, Thöni M, Trommsdorff V (1982) Magmatism and metamorphism in the Ladakh Himalayas (the Indus-Tsangpo suture zone). Earth Planet Sci Lett 60(2):253–292

Huang C, Zhang Q, Liu F (1989) A preliminary study of paleovegetation and palaeoclimate in the later period of late Pleistocene in the Bangong co Lake region of Xigang. J Nat Resour 4:247–253

Imsong W, Bhattacharya F, Mishra RL, Phukan S (2017) Geomorphic evidence of late quaternary displacement of the Karakoram fault in Nubra and Shyok valleys, Ladakh Himalaya. Curr Sci 112:2295–2305

Ives JD (1986) Glacial Lake outburst floods and risk engineering in the Himalaya. ICIMOD Occasional Paper 5:42

Jade S, Bhatt BC, Yang Z, Bendick R, Gaur VK et al (2004) GPS measurements from the Ladakh Himalaya, India: preliminary tests of plate-like or continuous deformation in Tibet. Geol Soc Amer Bull 116:1385–1391

Jain AK, Manickavasagam RM, Singh S, Mukherjee S (2005) Himalayan collision zone: new perspectives-its tectonic evolution in a combined ductile shear zone and channel flow model. Himal Geol 26(1):1–18

Jain AK, Singh S (2008) Tectonics of the southern Asian plate margin along the Karakoram shear zone. Constraints from field observations and U-Pb Shrimp ages. Tectonophysics 451:186–205

Jain AK, Singh S (2009) Geology and tectonics of the southeastern Ladakh and Karakoram. Geological Society of India, Bangalore, India, p 179p

Juyal N (2014) Ladakh: the high-altitude Indian cold desert. In: Landscapes and landforms of India. Springer, Dordrecht, pp 115–124

Kohn MJ (2014) Himalayan metamorphism and its tectonic implications. Annu Rev Earth Planet Sci 42:381–419

Kotlia BS, Shukla UK, Bhalla MS, Mathur PD, Pant CC (1997) Quaternary fluviolacustrine deposits of the Lamayuru Basin, Ladakh Himalaya: preliminary multidisciplinary investigations. Geol Mag 134:807–812. https://doi.org/10.1017/S0016756897007826

Kotlia BS, Bhalla MS, Shah N, Rajagopalan G (1998) Palaeomagnetic results from the Pleistocene-Holocene lake deposits of Bhimtal and Bhowali (Kumaun Himalaya) and Lamayuru (Ladakh Himalaya) with reference to the reversal events. Journal Geological Society of India 51(1):7–20

Kumar A, Srivastava P (2017) The role of climate and tectonics in aggradation and incision of the Indus River in the Ladakh Himalaya during the late quaternary. Quat Res 87(3):363–385

Kumar A, Srivastava P, Meena NK (2017) Late Pleistocene aeolian activity in the cold desert of Ladakh: a record from sand ramps. Quat Int 443:13–28

Lal R, Saini HS, Pant NC, Mujtaba SAI (2019) Tectonics induced switching of provenance during the late quaternary aggradation of the Indus River Valley, Ladakh, India. Geosci Front 10(1):285–297

Le Fort P (1975) Himalayas: the collided ranges. Present knowledge of the continental arc. Am J Sci 275A:1–44

Lisiecki LE, Raymo ME (2005) A Pliocene-Pleistocene stack of 57 globally distributed benthic δ18O records. Paleoceanography 20(1)

Liu Q (1993) Paleoclimate and chronological constraints on recent movements in Western Tibet: faults of Karakorum and Longmu co-Gozha co, pull-apart lakes of Longmu co and Sumxi co. Ph.D. thesis, Univ, Paris VII, Paris

Majeed U, Rashid I, Sattar A, Allen S, Stoffel M, Nüsser M, Schmidt S (2021) Recession of Gya glacier and the 2014 glacial lake outburst flood in the trans-Himalayan region of Ladakh, India. Sci Total Environ 756:144008

Moga J, Kohan B, Csullog G, Strat D (2019) The geomorphological survey of the pseudo karst in the area of Lamayuru Paleolake badlands (moon land), Ladakh, region, India. Geografia Fisicae Dinamica Quaternaria 42(1):87–98

Molnar P, Tapponnier P (1978) Active tectonics of Tibet. J Geophys Res Solid Earth 83(B11):5361–5375

Mujtaba SAI, Lal R, Saini HS, Kumar P, Pant NC (2017) Formation and breaching of two palaeolakes around Leh, Indus valley, during the late quaternary. Geol Soc Lond, Spec Publ 462(1):23–34

Munack H, Korup O, Resentini A, Limonta M, Garzanti E, Blothe JH, Scherler D, Wittmann H, Kubik PW (2014) Postglacial denudation of western Tibetan plateau margin outpaced by long-term exhumation. Bull Geol Soc Am 126:1580–1594. https://doi.org/10.1130/B30979.1

Nag D, Phartiyal B (2015) Climatic variations and geomorphology of the Indus River valley, between Nimo and Batalik, Ladakh (NW trans Himalayas) during late quaternary. Quaternary International, 371, pp. 87–101

Nag D, Phartiyal B (2015) Climatic variations and geomorphology of the Indus River Valley, between Spituk and Batalik, Ladakh (NW trans Himalayas) in late quaternary. Quatern Int 371:87–101

Norin E (1925) Preliminary notes on the late quaternary glaciation of the North-Western Himalaya. Geogr Ann 7:165–194

Owen LA, Derbyshire E, Fort M (1998) The Quaternary glacial history of the Himalaya. Journal of Quaternary Science 13:91–120

Owen LA, Caffee MW (2006) Terrestrial cosmogenic nuclide surface exposure dating of the oldest glacial successions in the Himalayan orogen: Ladakh range, northern India. Geol Soc Am Bullet 118:383–392

Owen LA, Finkel RC, Caffee MW (2002) A note on the extent of glaciation throughout the Himalaya during the global last glacial maximum. Quat Sci Rev 21:147–157

Pant, R.K., Phadtare, N.R., Chamyal, L.S. And Juyal, N., 2005: Quaternary deposits in Ladakh and Karakoram Himalaya: a treasure trove of the palaeoclimate records. Curr Sci, 88(11), pp. 1789–1798

Phartiyal B, Sharma A (2009) Soft sediment deformation structures in the late quaternary sediments of Ladakh: evidence for multiple phases of seismic tremors in the north western Himalayan region, India. J Asian Earth Sci 34:761–770

Phartiyal B, Sharma A, Kothari GC (2013) Damming of river Indus during late quaternary in Ladakh region of trans-Himalaya, NW India: implications to Lake formation-climate and tectonics. Chinese Sci Bull 58(I):142–155

Phartiyal B, Sharma A, Upadhyay R, Ram-Awatar, Sinha AK (2005) Quaternary geology, tectonics and distribution of palaeo- and present fluvio/glacio lacustrine deposits in Ladakh, NW Indian Himalaya—a study based on field observations. Geomorphology 65:241–256

Phartiyal B, Singh R, Kothyari GC (2015) Late-quaternary geomorphic scenario due to changing depositional regimes in the Tangtse Valley, trans-Himalaya, NW India. Palaeogeogr Palaeoclimatol Palaeoecol 422:11–24

Phillip G, Mazari RK (2000) Shrinking Lake basins in the proximity of the Indus suture zone of northwestern Himalaya: a case study of Tso Kar and Startsapuk Tso, using IRS 1C data. Int J Remote Sens 21:2973–2984

Phillips RJ (2008) Geological map of the Karakoram fault zone, eastern Karakoram, Ladakh, NW Himalaya. J Maps 4(1):21–37

Phillips RJ, Parrish RR, Searle MP (2004) Age constraints on ductile deformation and long-term slip rates along the Karakoram fault zone, Ladakh. Earth Planet Sci Lett 226(3–4):305–319

Phillips RJ, Searle MP (2007) Macrostructural and microstructural architecture of the Karakoram fault: relationship between magmatism and strike-slip faulting. Tectonics 26(3)

Qayyum M, Lawrence RD, Niem AR (1997) Discovery of the palaeo-Indus delta-fan complex. Geol Soc London 154(5):753–756

Rathour R, Gupta J, Mishra A, Rajeev AC, Dupont CL, Thakur IS (2020) A comparative metagenomic study reveals microbial diversity and their role in the biogeochemical cycling of Pangong lake. Sci Total Environ 731:139074

Ravikant V (2006) Utility of Rb-Sr geochronology in constraining Miocene and cretaceous events in the eastern Karakoram, Ladakh, India. J Asian Earth Sci 27:534–543

Reynolds JM (1998) High-altitude glacial lake hazard assessment and mitigation: a Himalayan perspective. In: Maund JG, Eddleston M (eds) Geohazards in engineering geology. Geological Society, vol 15. Engineering Geology Special Publications, London, pp 25–34

Richardson SD, Reynolds JM (2000) An overview of glacial hazards in the Himalayas. Elsevier Science Ltd and INQUA, Quat Int 65(66):31–47

Sachan HK, Mukherjee BK, Ogasawara Y, Maruyama S, Ishida H, Atsumi MUKO, Yoshioka N (2004) Discovery of coesite from Indus suture zone (ISZ), Ladakh, India. European J Mineral 16(2):235–240

Sangode SJ, Bagati TN (1995) Tectono-climatic signatures in higher Himalayan lakes: a palaeomagnetic rock magnetic approach in the lacustrine sediments of Lamayuru, Ladakh, India. J Himal Geol 6:51–60

Sangode SJ, Meshram DC, Rawat SUMAN, Suresh N, Srivastava P (2017) Sedimentary and geomorphic observations along Pangong strand of the Karakorum fault (NW Himalaya) depicting Holocene uplift. Himal Geol 38(2):111

Sangode SJ, Phadtare NR, Meshram DC, Rawat S, Suresh N (2011) A record of Lake outburst in the Indus Valley of Ladakh Himalaya, India. Curr Sci 102(9):1712–1718

Sangode SJ, Rawat S, Meshram DC, Phadtare NR, Suresh N (2013) Integrated mineral magnetic and lithologic studies to delineate dynamic modes of depositional conditions in the Leh Valley Basin, Ladakh Himalaya. J Geol Soc India 82:107–120

Sant DA, Wadhwan SK, Ganjoo RK, Basavaiah N, Sukumaran P, Bhattacharya AS (2011) Linkage of paraglacial processes from last glacial to recent inferred from Spituk sequence, Leh valley, Ladakh Himalaya. J Geol Soc India 78:147–156

Schmidt S, Nüsser M, Baghel R, Dame J (2020) Cryosphere hazards in Ladakh: the 2014 Gya glacial lake outburst flood and its implications for risk assessment. Nat Hazards 104(3):2071–2095

Searle MP (1996) Cooling history, erosion, exhumation, and kinematics of the Himalaya-Karakoram-Tibet orogenic belt. In: Yin A, Harrison TM (eds) The tectonic evolution of Asia. Cambridge University Press, New York, pp 110–137

Searle MP (2015) Mountain building, tectonic evolution, rheology, and crustal flow in the Himalaya, Karakoram, and Tibet. In: Schubert G (ed) Treatise on geophysics, vol 6. Elsevier, Oxford, pp 469–511

Searle MP, Owen LA (1999) The evolution of the Indus River in relation to topographic uplift, climate and geology of western Tibet, the trans-Himalayan and high-Himalayan range. In: The Indus River: biodiversity, resources, humankind. Oxford University Press, Delhi, p 230

Searle MP, Phillips RJ (2007) Relationships between right-lateral shear along the Karakoram fault and metamorphism, magmatism, exhumation and uplift: evidence from the K2–Gasherbrum–Pangong ranges, North Pakistan and Ladakh. J Geol Soc 164(2):439–450

Searle MP, Tirrul R (1991) Structural and thermal evolution of the Karakoram crust. J Geol Soc Lond 148:65–82

Searle MP, Weinberg RF, Dunlap WJ (1998) Transpressional tectonics along the Karakoram fault zone, northern Ladakh: constraints on Tibetan extrusion. Geol Soc Lond, Spec Publ 135(1):307–326

Searle MP, Windley BF, Coward MP, Cooper DJ, Rex AJ (1988) The closing of Tethys and the tectonics of the Himalaya. Geol Soc Am Bull 98(6):678–701

Searle MP, Cottle JM, Streule MJ, Waters DJ (2009) Crustal melt granites and migmatites along the Himalaya: melt source, segregation, transport and granite emplacement mechanisms. Earth and Environmental Science Transactions of the Royal Society of Edinburgh 100(1–2):219–233

Sharma KK (1992) Petrology, geochemistry and geochronology of the Ladakh batholith and its role in the evolution of Ladakh magmatic arc. In: Sharma KK (ed) Geology and geodynamic evolution of the Himalayan collision zone, vol 17. Physics and Chemistry of the Earth, pp 173–194

Shi Y, Yu G, Liu X, Li B, Yao T (2001) Reconstruction of the 30–40 ka BP enhanced Indian monsoon climate based on geological records from the Tibetan plateau. Palaeogeogr Palaeoclimatol Palaeoecol 169(1–2):69–83

Shroder JF, Bishop MP (1999) Indus to the sea: evolution of the system and Himalayan geomorphology. In: Meadows A, Meadows PS (eds) The Indus River: biodiversity, resources, Humankind. Oxford University Press, Delhi, p 248

Shukla DP, Dubey CS, Singh N (2012) Neotectonic activity and the origin of Tso Morari Lake using remote sensing and digital elevation model (DEM) derivative techniques. Geocarto Int 27(3):249–262

Shukla UK, Kotlia BS, Mathur PD (2002) Sedimentation pattern in a trans-Himalaya quaternary Lake at Lamayuru (Ladakh), India. Sediment Geol 148:405–424

Shukla AD, Sharma S, Rana N, Bisht P, Juyal N (2020) Optical chronology and climatic implication of glacial advances from the southern Ladakh range, NW Himalaya, India. Palaeogeogr Palaeoclimatol Palaeoecol 539:109505

Srivastava P, Kumar A, Chaudhary S, Meena N, Sundriyal YP, Rawat S, Rana N, Perumal RJ, Bisht P, Sharma D, Agnihotri R (2017) Paleofloods records in Himalaya. Geomorphology 284:17–30

Srivastava P, Kumar A, Singh R, Deepak O, Kumar AM, Ray Y, Jayangondaperumal R, Phartiyal B, Chahal P, Sharma P, Ghosh R (2020) Rapid Lake level fall in Pangong Tso (lake) in Ladakh, NW Himalaya: a response of late Holocene aridity. Curr Sci 119(2):219–231

Suri K (2018) Understanding historical, cultural and religious frameworks of mountain communities and disasters in Nubra valley of Ladakh. Int J Dis Risk Reduct 31:504–513

Thakur VC, Misra DK (1984) Tectonic framework of the Indus and Shyok suture zones in eastern Ladakh, northwest Himalaya. Tectonophysics 101(3–4):207–220

Thakur VC, Rawat BS (1992) Geologic map of Western Himalaya, scale 1:1,000,000. Wadia Institute of Himalayan Geology, Dehradun, India

Thanh NX, Sajeev K, Itaya (2011) Multiple garnet growth in garnet–kyanite–staurolite gneiss, Pangong metamorphic complex, Ladakh Himalaya: new constraints on tectonic setting. Lithos 127:552–563

Üner S, Alırız MG, Özsayın E, Selçuk AS, Karabıyıkoğlu M (2017) Earthquake induced sedimentary structures (Seismites): Geoconservation and promotion as geological heritage (Lake Van-Turkey). Geoheritage 9(2):133–139

Üner S, Özsayın E, Selçuk AS (2019) Seismites as an indicator for determination of earthquake recurrence interval: a case study from Erciş fault (eastern Anatolia-Turkey). Tectonophysics 766:167–178

Upadhyay R (2001) Seismically-induced soft-sediment deformational structures around Khalsar in the Shyok Valley, northern Ladakh and eastern Karakoram, India. Curr Sci:600–604

Upadhyay R (2003) Earthquake induced soft sediment deformation in the lower Shyok river valley, northern Ladakh, India. J Asian Earth Sci 21(4):413–421

Wright TJ, Parsons B, England PC, Fielding EJ (2004) In SAR observations of low slip rates on the major faults of Western Tibet. Science 305(5681):236–239

Ziegler AD, Cantarero SI, Wasson RJ, Srivastava P, Spalzin S, Chow WT, Gillen J (2016) A clear and present danger: Ladakh's increasing vulnerability to flash floods and debris flows. Hydrol Process 30(22):4214–4223

Volcanology of Unique Geoheritage Sites in Deccan Region of Western India

Anshuman Misra and K. S. Misra

Abstract Physical volcanic features associated with lava tube, channels, tunnels and effusive centers have been utilized by the ancient artisans for creating some of the magnificent geoheritage sites in the Deccan region of Western India. This region is covered by thick, black, dull luster basalt units with inherent presence of cooling cracks, iron and ferromagnesian minerals. It shatters during chiseling and accelerated weathering gives permanent brownish stains. Thermal baking and induration in zones close to lava tubes, channels, tunnels and effusive centers, has hardened the rocks, sealed the cracks, amalgamated iron and ferromagnesian minerals and attained vibrant shades of red to become suitable for excavations and sculpturing. Remnants of lava tubes and channels, composed of thermally baked, indurated and reddened basalt were recognized as an excellent rock for carving the statues and idols as well as exceptional dimension stone. Degree of baking and induration gets reduced with increasing distance and is also utilized in deciding sculpturing pattern. Hollow cylindrical tubes emptied by draining out of lava in high gradient areas, were found suitable for innumerable cave temples. Saptashrungi temple, at the portal of lava tube; Sita Mata cave, Nasik; Bhartahari Cave, Ujjain; Pataleshwar and Baneshwar temples, Pune; are examples of lava tunnels. Innumerable generations with different religious beliefs, such as Buddhists and Hindus, have worked in tandems for centuries to create Ajanta, Ellora, Elephanta, Kanehri, Karla, Bhaja cave complexes; in favorable regions. Frequent attacks by invaders were resisted; however, large-scale destruction in several cave complexes were repaired by the matching baked and indurated rocks.

Keywords Volcanology · Lava tubes · Channels and tunnels · Thermal induration · Geoheritage sites · Deccan region · Western India

A. Misra (✉)
Department of Geology, Doon University, Dehradun, Uttarakhand, India

K. S. Misra
Department of Petroleum Engineering and Earth Sciences, The University of Petroleum and Energy Studies, Dehradun, Uttarakhand, India

S. C. Tripathi et al. (eds.), *Geoconservation and Geotourism Potential of India*, Society of Earth Scientists Series, https://doi.org/10.1007/978-3-031-81017-6_4

1 Introduction

This article illustrates the mind-boggling ingenuity of ancient artisans, who utilized very complex geological details and skillfully utilized them in excavating and sculpturing unparalleled time-defying monuments. The Deccan region of western India is covered by approximately 60 to 65 million-year-old volcanic sequences, composed mainly of basalt and minor associated rocks. As such, the basalt is not suitable for excavation or sculpturing because of the inherent presence of cracks, even at micro level, developed due to volume reduction during cooling and thus shatters during chiseling. Furthermore, basalts is black, unattractive dull luster and invariably have iron and several other ferromagnesian minerals. It chemically weathers fast in hot, humid climatic conditions and gives repulsive permanent brownish stains. During mechanical weathering, characteristic spheroids, similar to the onion shells, develop with differing degrees of friability, rendering them useless for carving. Thus the vast terrain of basalts does not qualify for in-situ rock-cut monuments and attractive idols and statues to last for centuries.

In the terrain of several million sq. km. in western India, the ancient artisans have identified rare thermally baked zones and indurated in the vicinity of volcanoes and volcanic fields with emanating lava channels and tubes. Due to this secondary heating, the cracks were sealed, iron minerals got amalgamated and these rocks acquired vibrant shades of red. They also attained hardness close to 6 on Moha's scale, became quite suitable for permanent excavations, sculpturing, polishing as well as very durable dimension stone. Resistance to chemical and mechanical weathering made these rocks best suited for excavation of Ellora, Ajanta, Pitalkhora, Elephanta, Kanehri, Karla, Bhaja cave complexes; invincible forts such as Deogiri close to Daulatabad. The degree of baking and induration is also utilized skillfully in deciding the sculpturing pattern and utilization as dimension stone. An integrated arterial system, comprising lava tubes and channels through which molten basaltic lava has flowed to long distances, has been presented by Misra et al. (2001) and Misra (2002a). Effusive center at Asolia Dongar was identified from the emergence of lava channel (Misra 2002b). Remnants of lava tubes and channels are identified by Misra (2005) in more than three hundred and fifty places in Deccan volcanic terrain. Effusion of volcanics along the major rift and grabens has been illustrated earlier (Misra 2016).

Along with baking and induration, the ancient artisans have also utilized the disposition of rock units; physical volcanic features related to lava channels, tubes, tunnels have been skillfully utilized. New parameter to identify covered effusive centers by radiating pattern of lava channels is described in this study. A set of two prominent channels are seen emerging both on the western as well as on northern sides of Aurangabad plateau. It has also emerged that multi-central effusion has taken place and local volcanic stratigraphy.

Innumerable generations continued the activity with geometrical precision. Artisans with different religious beliefs, such as Buddhists and Hindus have worked in tandems for centuries to achieve insurmountable goals. Frequent attacks by the invaders were resisted; however, large-scale destruction in several cave complexes is evident and has also left very many caves unfinished.

The present study significantly contributes to repairing and preserving geoheritage sites, statues, and idols. In many sites the statues and idols have been damaged by the invaders and at other places there is deterioration due to weathering and erosional processes over centuries. The repairs are generally done by using cement and concrete. The study suggests the use same geological indurated basalt to repair astatically or complete replacement. The presence of lava tubes with a diameter of about 100 m and lava channels more than 300 m wide indicates the formidable availability of indurated and reddened basalt in Deccan volcanic terrain.

2 Disposition of Volcanic Units and Geoheritage Sites

Horizontal disposition of successive volcanic flows can be seen in the entire terrain; however, the most impressive sections are exposed along the Sahyadri Mountain Ranges (Fig. 1). It can be seen that alternate dark and light tones are due to the vegetation cover. The softer units are thickly vegetated and have slightly gentler slopes with soil cover. On the other hand, the hard bands are more massive, resistant and consequently gave rise to steeper slopes without soil cover.

Fig. 1 Panoramic view showing horizontally disposed successive volcanic flows, exposed in the Lohagadh Fort section close to Kamshet Township

Fig. 2 Cave complexes have utilized horizontally disposed successive volcanic flows. Massive hard bands jutting out of the vertical section can be seen utilized as roof tops

This varying hardness of individual flows is very well utilized in extensive excavations. The harder ones, jutting out in the vertical section, are selected for rooftops. In contrast, less hard bands are excavated for vast meditation halls (Fig. 2). Beautifully carved arched entrances to support roofs are selected according to strength, weathering and fracture pattern.

The ancient artisans were at their best in selecting excavation sites. Hindus made temples and Buddhists excavated large meditation halls (Vihars), primarily based on geomorphic considerations. Magnificent cave complexes at Ellora, Ajanta, Elephanta, Kanheri, Karla and Bhaja, excavation is largely done along a particular contour on cliff section. These caves have elaborate elongated meditation halls with intricately carved pillars and arches. The beams are sculptured in flows having higher bulk strength and are supported by pillars, while arches support the weaker horizons. Different types of carved lamp support, provided favourable lighting for meditation (Fig. 3).

Fig. 3 Horizontal disposition of volcanic units is utilized in excavation of vast vihars. Layering can be seen across the meditation hall. The units have differential erodibility, are used with great ingenuity. The beams are carved in comparatively hard units, while softer units are protected by arches to avoid rock fall from the roof or sides and eventual collapse. Layering can be seen in all the pillars across the meditation hall

3 Lava Channels, Tubes, Tunnels and Geoheritage Sites

The existence of multi-tier system of flowing very hot lava through channels, tubes and tunnels has been a great source of heat and has not only baked the levee portions but also transferred heat to the surrounding rocks. The intensity of this heat transfer can be judged by the degree of of baking and induration due to the flow of very hot lava for long duration. In field, they are recognized by positive relief of levee portions, sinuous shapes and being composed of reddened fragmentary volcanic material (Fig. 4).

The most common remnants are the basal portions of the lava channels, followed by the levee portions and the rarest are the cross sections. The best cross-section is seen close to the Sikri Village in Ahmednagar district (Fig. 5). Upward converging columnar joints are seen along the other sections in the continuity of the lava channel. The channel is approximately 30 m across, however, from the disposition of columnar joints the diameter up to 250 m is also ascertained.

An excellent utilization of physical volcanology is seen in site selection of Saptashrungi temple complex, located in the portal portion of a large lava tube. The cliff section exposes hardened and reddened basalt around the portal (Fig. 6).

Fig. 4 Panoramic view of remnant of lava channel in Gunjale-Vambori area. Nearly parallel levee portions having sinuous shape and positive relief can be seen

Fig. 5 Field photograph shows cross-section of a lava tube near Sikri village in Ahmednagar district, Maharashtra. Upward converging columnar joints can well be seen. The tube filling material is fragmentary and easily erodible. The encasing rock is very hard, reddened indurated basalt

Fig. 6 Saptashrungi temple complex, located in the portal portion of lava tube. Indurated reddened rock is exposed forms the tube

Fig. 7 Panoramic view of the Saptashrungi cliff portion. At the base the township is located; the straight upward structure is escalator, while the gentle meanders are steps to the temple

A 3-dimensional view generated from the satellite data has shown that hard basalt flow stands out as a sharp-crested ridge. The portal of the lava tube is located within the hard lava flow (Fig. 7).

Fig. 8 Lava tunnel near Uksan Village, close to Kamshet Township on old Mumbai-Pune Highway. The base is nearly horizontal, while the roof is arched. Chiseling to make sitting place on the margin can also be seen

There are areas where a number of lava tunnels have been identfied in Deccan volcanic region, Perhaps the best among the so far located, are Uksan and Pale tunnels near Kamshet on Mumbai-Pune old highway (Fig. 8).

Pashan-Baner hill transecting the Pune city is found to have lava tunnel at the eastern end. This tunnel is suitably excavated in indurated and reddened basalt, for Baneshwar temple, several centuries ago (Fig. 9).

The present study suggests that the extensive Deccan Plateau is composed of several smaller plateaus. Among them the most prominent with a number of geoheritage sites, is the Aurangabad plateau. A number radially emanating remnants are identified. The most important one is starting from the north of Kailash temple and proceeds right up to Grishneshwar temple in the west (Fig. 10).

The entire temple with meticulously carved, large pillars, columns, beams, top and floor are made of reddened indurated basalt. Impeccable rocks have been selected for the construction. The rock takes excellent polishing and attains an exceptionally brilliant look and used with a very high degree of symmetry and intricate carving done centuries ago is still flawless, without any maintenance or painting. The artisans were not only excellent in chisel and hammer but also in polishing. It is unclear how such excellent rock polishing was done at that time without modern abrasives (Fig. 11). In Grishneshwar temple complex, the ancients have graded

Fig. 9 Baneshwar cave temple on ridge close to Baner Village in Pune. Fairly well indurated rocks have been excavated to make this geoheritage site centuries ago

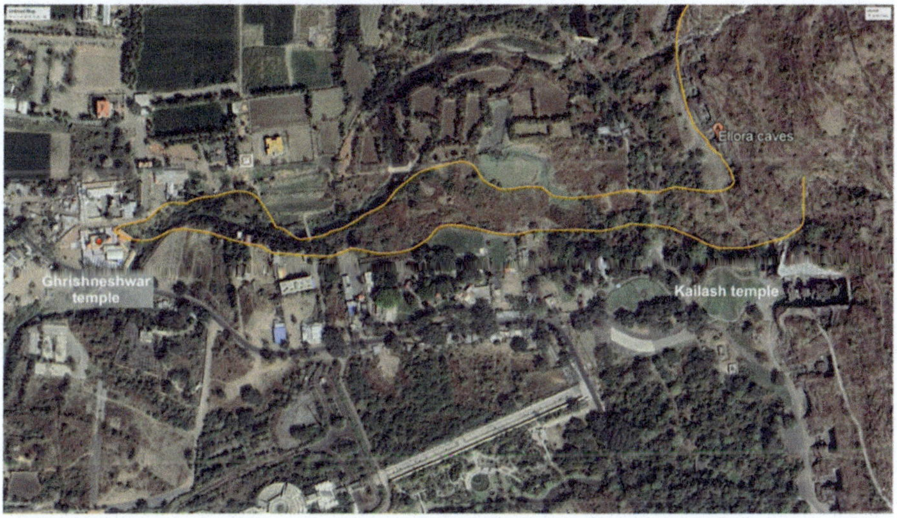

Fig. 10 Satellite picture shows the levee portions of lava channel from Ellora cave complex, a little north of Kailash temple extending to Grishneshwar temple

and utilized skilfully. Dimension stones with white vesicles and pipe amygdules are used for pavement and boundary walls.

Highly indurated and reddened basalt with scattered, pipe, and horizontally layered amygdules, filled with silica and zeolites are very commonly used in carving the idols (Fig. 12), flooring and walls in the temple complex.

Fig. 11 Intricately carved pillar in the Grishneshwar temple, located to the west of Ellora cave complex. Large pillars, beams, top and floor are made of reddened and highly indurated basalt

Another quite prominent remnant of lava channel is seen emanating from the Aurangabad plateau, towards west right up to Tisgaon locality (Fig. 13).

Remnants of lava channels emanating from the northern escarpment, are also mapped. A very complex pattern of lava channels existed in the area, approximately 2 km east of Ghusur village. Other smaller distributary channels are seen from the trunk channel (Fig. 14).

Fig. 12 Idol of Nandi, carved in highly indurated reddened basalt with white silica and zeolite filled vesicles

Fig. 13 Remnant of lava channel emerging from the Aurangabad plateau to Tisgaon locality in the west

Fig. 14 Complex pattern of remnants of lava channels emanating from the Aurangabad plateau towards north can be seen

4 Effusive Centers and Geoheritage Sites

Close association between effusive centers and geoheritage sites is seen at a number of places. From geological consideration, our study suggests that the Ellora cave complex is located in the vicinity of a huge covered effusive centre with a complex pattern of lava tubes and channels. Other evidences, such as highly vasicular and,

Fig. 15 Satellite view of Deogiri Fort, close to Daulatabad Town. The towering hill has a water filled protecting canal and vertical cliff all around. A number of lava channels emanate radially from the base of the hill

amygdular nature, horizontal zeolite layering, ropy lava, baked and reddened basalt, indicate proximity with the effusive centre. Perhaps the best example of the utilization of Geology and Geomorphology in site selection, is the invincible Deogiri Fort close to Daulatabad town Fig. 15). The horizontal contour is utilized for excavation of protecting canal all along the base of hill.

Fort located on the top of towering Deogiri Hill, comprising alternating simple, hard and compact 'aa' type basalt flows and form inassesable vertical cliffs. The fort wall is impenetrable, as it is made of either the natural hard rock or the large rock blocks. Enterances and steps within the fort are excavated in auch a manner that not more than one person can enter and climb the steps. A number of lava channels, radiating out from the hill are identified (Fig. 16). These channels have characteristic levee portions composed of aglutinated fragmentary material and nodular basalt in the central portion.

Fig. 16 Deogiri Hill close to Daulatabad, composed of alternate resistant (forming vertical cliffs) and less resistant (forming steep slope) units. A remnant of lava channel can be seen coming out of the hill

5 Dimension Stones and Geoheritage Sites

Dimensions extracted from the remnants and from the vicinity of lava tubes/channels and effusive centers have been utilized for centuries in construction of temples (Ghrishneshwar, Khondoba and vithal-Ruhmini);, tombs (Bibi-ka-Makbara), churches, memorials and heritage buildings, forts (Deogiri, Daulatabad, Singhagarh) to name a few. Degree of induration has also been skillfully utilized. Patricianly baked rock is used in construction of fort walls (Fig. 17).

Dressed blocks of highly indurated basalt are most favored in construction of modern buildings (Fig. 18).

In far off Malwa volcanic region, Mangalnath and several smaller temples in Ujjain; and around Indore city are constructed by using the indurated basalt units. Similarly, in Shri Vitthal-Rukmani temple in Pune, the precious reddened basalt is used in constructing the main temple (Fig. 19) and surrounding walls with less indurated basalt.

Apart frombaking and induration, two other volcanic processes have played important role in formation of geoheritage rock. They are composition and texture. Trachyte an alkaline varient of basalt, is used in construction of the famous landmark memorial of Mumbai, The Gateway of India. This towering structure is not affected by Monsoonsas well assalt laden winds for over a century. Similarly the Palace at Jawar, constructes of synite rock in the otherwise basalt terrain. Texture of

Fig. 17 Deogiri Fort wall made of partially baked and differently textured basalt

Fig. 18 Dressed blocks of highly indurated basalt are used very frequently in construction of buildings. No cleaning and painting is required for maintenance of these buildings

rocks has also affected the rock selection for particular type of carving. Extremely coarse agglomeratic rock, embedded with very large chunks of rock in Kanehri Cave region is used for excavation of caves (Fig. 20). While exceedingly fine rextured agglomerate rock is used for very fine and intrcete sculpturing in Elephanta Cave complex (Fig. 21).

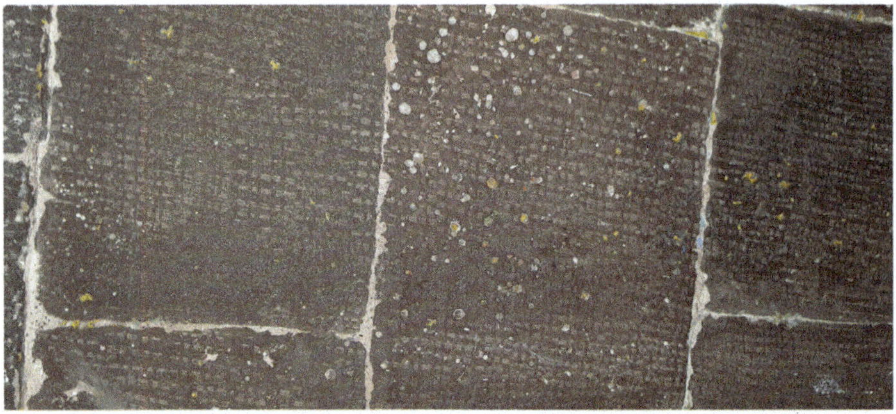

Fig. 19 Blocks of indurated basalt used in construction of walls of Shri Vallitthal-Rukmini temple in Pune. Vesicles are filled by white secondary silica and zeolites can be seen. Tiny squares are chiseled on rock to make it attractive

Fig. 20 The volcanic rock has large boulders of different shapes. The elongated ones are horizontally placed. Upward fining can also be seen

Fig. 21 Exceedingly fine grained agglomerate rock is used for intricate carving in Elephanta Cave Complex

6 Summary and Conclusions

Natural basalt rock located in the proximity of lava tube and channels as a well as effusive centers, is baked and indurated, has been utilized by the ancient artisans for creating some of the magnificent geoheritage sites. During thermal baking the rock has hardened recrystallized; joints, fractures are sealed and acquired vibrant red color. Degree of baking and induration is also utilized in deciding sculpturing pattern with geometrical precision. Grishneshwar temple, Ajanta and Ellora cave complexes are excavated in highly baked levee portions of lava tubes and channels, and proximity with the effusive centers. Normal basalt rock weathers quite fast as it has iron minerals and gives rusty stains only within few years and thus it is not preferred for exposed surfaces. In contrast, the indurated basalt does not show any weathering effects even after several centuries.

The present study significantly contributes to repairing and preserving geoheritage sites, statues, and idols. In many sites the statues and idols have been damaged by the invaders and at other places there is deterioration due to weathering and erosional processes over centuries. The repairs are generally done by using cement and concrete. The study suggests the use same geological indurated basalt to repair astatically or complete replacement.

Most common volcanic rock, the basalt is black and has an unattractive dull luster. Presence of cooling cracks renders it useless for sculpturing as well as dimension stone. Furthermore, basalts invariably have iron and several other ferro-magnesian minerals, which chemically weather fast in hot, humid climatic conditions and give repulsive permanent brownish stains. Due to baking, the cracks get sealed, iron minerals got amalgamated and rocks acquired vibrant shades of red. In ideal cases the rock attains hardness close to mineral orthoclase (6 on Moha's scale), became quite suitable for permanent excavations, sculpturing, polishing as well as very durable dimension stone.

Very complex geological details are utilized by ancient artisans for centuries for excavating and sculpturing unparalleled time-defying monuments. Processes involved with volcanic activity such as baking and induration have modified basalt rock to most magnificent monumental stone. The ancient artisans realized these properties of different degrees of baking and induration and color, to accordingly select different types of artwork. The highest degree of induration seen in western India exists in the Aurangabad region and explains the presence of many monumental sites. Highly indurated rock is also most preferred for intricate carving of the masterpiece Grishneshwar temple. In the same temple complex, they have graded and utilized those with white vesicles and pipe amygdules for pavement and boundary walls.

The excavations in Ajanta, in Kailesh temple complex Ellora cave are enormous and historical evidences suggest that sculpturing continued for several centuries. Obviously, the skills for identifying induration, its degree and changes in textural characteristics were passed to successive generations. Remnants of lava tubes and channels are best exposed in areas where the escarpments have recently receded. Remnant rocks can be utilized to restore monumental geoheritage sites. The vibrant red colored basalt also takes excellent polish and attains an exceptionally brilliant look is generously used inside Grishneshwar temple. It is unclear how such excellent rock polishing was done at that time without modern abrasives. This study of geoheritage sites, in light of physical volcanology, has led the authors to develop new criteria for identifying thermally baked and indurated areas by studying the ancient temple complexes.

Data Availability

All data or data source incorporated are mentioned in the article.

Acknowledgements We wish to profusely thank Dr. Satish C. Tripathi, Former Deputy Director General, Geological Survey of India; Honorary Editor, Earth Science India and General Secretary, The Society of Earth Scientists; for inviting us to contribute this article to forthcoming book on Geoheritage. Prof. N C Pant, Department of Geology, Delhi University gave valuable suggestions of great help to significantly improve the manuscript. Landsat images from Google Earth, used during the study are thankfully acknowledged.

Funding No funding is received to do this research.

Ethics Declarations Competing Interests

The authors declare no competing interests. There are no financial or non-financial interests that are directly or indirectly related to the work submitted for publication.

References

Misra KS, Sable AB, Bhutani R (2001) Lava channels and tubes in the western part of the Deccan Volcanic Province. Geological Survey of India, Special Publication, pp 503–510

Misra KS (2002a) Arterial system of lava tubes and channels within Deccan Volcanics of western India. J Geol Soc India:115–124

Misra KS (2002b) Arterial system of lava tubs and channels within Deccan Volcanics of western India, Discussion. J Geol Soc India 59:595–600

Misra KS (2005) Distribution pattern, age, duration and mode of eruption of Deccan and associated Volcanics, Gondwana geological magazine. Special 8:53–60

Misra KS (2016) Extensional tectonics, rifting, formation of Sedimentary Basins, Cretaceous Volcanism, Emplacement of Dyke Swarms and Development of Hydrocarbon Pools, Case Studies from Peninsular India and Indian Ocean Region India, Acta Geologica Sinica English Edition, 90 supp. 1, pp. 59–60

Geoheritage Sites of the Palaeo-Saraswati River in Northwestern India in the Context of Geotourism

H. S. Saini, Naresh Chandra Pant, Apurva Aalok, and Abul Amir Khan

Abstract Geosites represent a large spectrum of features and sites of scientific significance and/or spectacular appearance. Geosites of archaeological significance awoke great curiosity in recent time and offer high potential of geotourism and research. Recent studies in the northwestern Indian plains have brought out stronger scientific evidence for (i) the existence of a Himalayan glacier-fed river now present as palaeochannel/s of characteristic hydrological traits (ii) a significantly large hydrological regime (iii) existence of now extinct Harappan age mounds association and (iv) composite geological-archaeological mounds indicating human habitation significantly older than the Harappan civilization. These geoheritage sites have great potential for geotourism but require immediate protection.

Keywords Geoheritage · Saraswati River · Harappan age · Palaeochannel · Palaeoclimate

1 Introduction

Unique looking landforms and rock outcrops preserving spectacular products of geological history or evidence of past-climatic processes are generally called as geosites or geomorphosites. These often lead to a sense of awe and wonderment and

H. S. Saini (✉)
Geological Survey of India, Faridabad, Haryana, India

N. C. Pant
National Centre for Polar and Ocean Research, Goa, India

A. Aalok
Altinok Consulting Engineering, Delhi, India

A. A. Khan
Amity Centre for Air Pollution Control (ACAPC), Amity University Gurugram, Manesar, Haryana, India

S. C. Tripathi et al. (eds.), *Geoconservation and Geotourism Potential of India*, Society of Earth Scientists Series, https://doi.org/10.1007/978-3-031-81017-6_5

become geotouristic sites, where people make special travel and associated logistic efforts to visit. There are numerous examples of such sites e.g. Niagara Falls in the USA, the barren "cold" desert landscape of Ladakh, Mt. Vesuvius in Italy and Arthur's seat (an extinct volcano) in Edinburgh and many others. Potential of these sites for geotourism, especially hitherto less explored rural geotourism (e.g. Mahato and Jana 2021), is yet to be realized in India. These georesources and their tourism potential is well recognized (e.g. Santangelo and Valente 2020 and the papers in that volume; Santangelo et al. 2020). UNESCO's "the Arouca declaration" (2012) attempts to resolve the duality of these sites as geographical or geological. Amongst these geosites, geoarchaeological sites are of special interest in context of geotourism as the natural geological/geocultural features gets interwoven with the evolution of the human societies as well as their decimation. Several studies in this context are available (e.g. Asrat and Zwoliński 2012 and the papers in that volume). These have been interestingly adopted for specific touristic groups (e.g. for cycling European tourists; Hose 2018).

Saraswati, a palaeo-river is a prominent surface and shallow-subsurface feature in India, which has religious, social and hydrological importance and, therefore, a geoheritage site of significance. It is also of great scientific curiosity because of its disappearance due to debatable reasons during the Holocene as well as its association with the Bronze age Harappan/Indus Civilization which flourished during 5.7–3.3 ka (Kenoyer 1998; Sahai 1999; Dikshit 2012) with an urban phase during ca. 2600/2500–1900 BC. Because of these, it finds mention for over 150 years in the scientific literature (Wilhelmy 1969; Oldham 1893) largely as a palaeochannel river whose segments were prominent in the northern parts of India from the Himalayas to the states of Haryana, Punjab and northern Rajasthan. In the remaining parts of desertic tracts of Rajasthan and finally in Gujrat, the traces of river courses are unclear (Yashpal 1980; Sood and Sahai 1983; Gupta et al. 2004; Bhadra et al. 2009; Mehdi et al. 2016). The geoheritage associated geotourism potential of the Harappan sites in the near Arabian sea coastal domain of Kachchh is well recorded and are indicated to be linked to the Saraswati palaeo-river (e.g. Thakkar et al. 2022).

Quaternary alluvial plains of Punjab and Haryana, south of the Himalayan front, constitute the most critical sector for the palaeo-Saraswati river research as it hosts some well-preserved archives of the past alluvial landforms which were not affected by massive desertification unlike the downstream reaches of the Saraswati in Rajasthan. Two types of features are prominent, a big loop of palaeochannel near Fatehabad, and numerous archeological mounds along, as well as, away from the loop of palaeo-river (Fig. 1). However, many of these features have been modified by agricultural activities in the last few decades. In the present work, in context of geoheritage and geotourism potential, we focus on three features on account of their significance in Holocene palaeoclimatic changes and societal evolution and disintegration of Harappan/Indus civilization. These sites/locales of interest are located on the northeastern fringes of the Thar Desert and alluvial plains south of the Siwalik front. In addition, a unique geological unit precipitated during Holoence as isolated lensoidal argillaceous limestone of lake origin also occurs in Haryana plains which

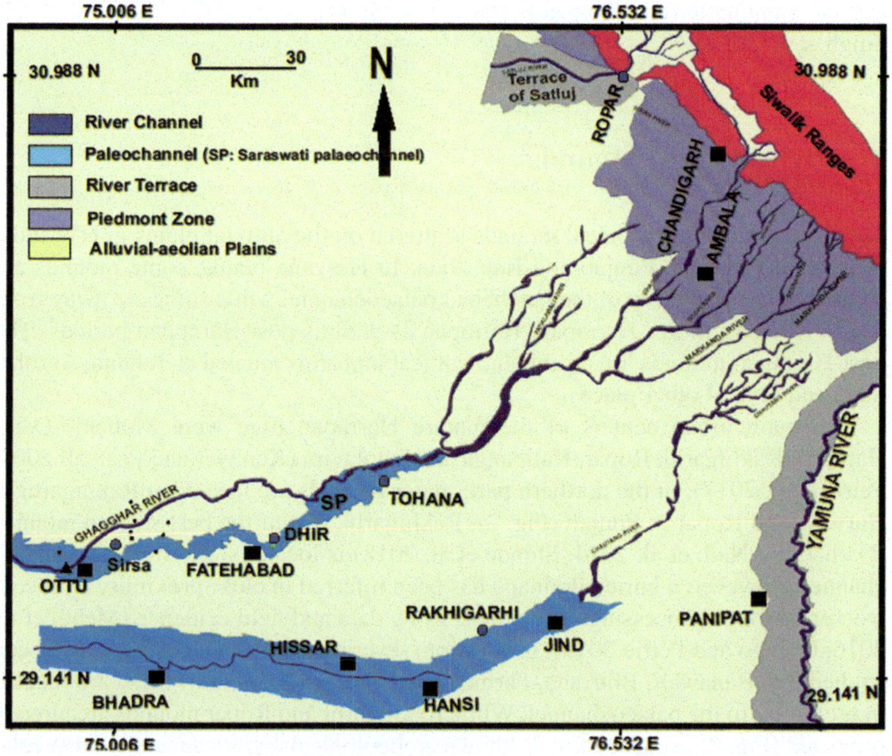

Fig. 1 Map of a part of NW Indian Plains and Himalayan front showing modern drainage of Ghagghar and Yamuna rivers and palaeochannels. The palaeochannel of Saraswati (SP) is most prominent around Fatehabad while southern palaeochannel along Jind- Hisasr- Bhadra is completely buried (Modified after Alok et al., 20,121)

marks a wet lake phase but the unit is now extremely scare due to extensive extraction.

Saraswati Palaeochannel network: This palaeochannel is 5–15 km wide zone starting from Tohana in north east to Ottu in south-west for >80 km length (Saini et al. 2020, Fig. 1). The zone stands out in the imagery as a high moisture zone characterized by silty lithology, fresh water bearing aquifers and low, nearly imperceptible depression that becomes apparent when flood water accumulates. It is more conspicuous near Fatehabad as a subtle lowland characterized by high ground moisture, clayey surface and fresh ground water (Saini and Mujtaba 2010). West of Fatehabad it is discernable in the form of a clayey, bumpy surface prone to water ponding during rains. This palaeochannel represents remnant of the Saraswati palaeo-river. A detailed network of this river system has been inferred by Mehdi et al. (2016) using remote sensing data. The river is believed to be a glacial-fed mighty river, which disappeared due to tectonic and climatic reasons (Valdiya 2013). A second view considers it a piedmont river also fed by a tributary of Sutlej and the Yamuna Rivers. (Tripathi et al. 2004; Giosan et al. 2012; Clift et al. 2012; Singh et al. 2017).

2 Archeological Mounds

There are many archeological mounds scattered on the alluvial plains in NW India in parts of Haryana, Punjab and Rajasthan. In Haryana plains, some mounds are located on the periphery of the Fatehabad palaeochannel while some are away from it. They belong to Pre-Harrapan, Harrapan as well as post-Harappan periods. The post-Harrapan mounds are prominent, conical lanforms located at Tohana, Agroha, Sirsa and several other places.

The main urban centers of the mature Harrapan time were Mohenjo-Daro, Harappa, Rakhigarhi, Ropar, Kalibanga and Dholavira (Kenoyer and Pearsall 2008; Petrie et al. 2017). In the northern part, major mounds are located at Rakhigarhi in Haryana and Ropar in Punjab (Fig. 2). Rakhigarhi, one of the best studied mounds (Nath 1998; Nath et al. 2014; Shinde et al. 2018) is located away from the palaeochannel. However; a buried drainage has been inferred in close proximity by recent workers through processing of remote sensing data and field evidence (Mehdi et al. 2016; Orengo and Petrie 2017). In addition, several smaller mounds have also been studied e.g. Banawali, Bhirrana, Farmana and Banawali. Most of these are located in proximity to the palaeochannel. While Rakhigarhi and Ropar mounds are already protected (Fig. 2) because of their known archeological significance, the other relatively lesser known and unknown mounds of also of great geo-archeological significance are either partially damaged or under threat. Some of these have yielded societal signatures as old as 10 ka, much older than the known age of Harrapan (Sarkar et al. 2016; Alok et al. 2021). In a multiproxy study, Alok et al. (2021) reestablish mineralogical, sedimentological and geochronological evidence for linking sediment sourcing of the channel from the Higher Himalaya. They also showed, for the first time, that the Saraswati palaeochannel was flooded as late as ~1500 years BP.

These signatures, impressions and archives of Sarswati palaeochannel and the archeological mounds are in different stages of obliteration, disfigurement and destruction because of uncontrolled anthropogenic activities (Fig. 3). The smaller mounds in the private fields are being cut and the material is used for filling purposes. There is a need of invoking a case for their preservation as record for future scientific, palaeoclimatic and palaeosocietal studies and geoeducation and geotourism for the general public. It is pertinent to mention that the state government of Haryana has initiated actions on rejuvenation of this palaeochannel.

1. **Fatehabad palaeochannel**: During 1983 to 1986, one of the authors (HSS) carried out Quaternary geological and geomorphological mapping of NW Haryana. At that time the palaeochannel zone was a perceptible, wet depression with clayey surface covered by around 2 m high grasses. There were only few patches of cultivation. The southern bank was discernable by a discontinuous scarp and

Fig. 2 Harrapan Archeological mounds, (**a**) Rakhigarhi, Haryana, (**b**) Ropar, Punjab, (**c**) Dhir, Haryana. Year of photography is given in brackets

the depression was flood prone during rains. In our repeat study during 2000–2003, the long grasses had disappeared, the channel was profusely under cultivation and the palaeobank scar was seen at few locations near Fatehabad, Hoshanga and Baisrana only (Saini and Mujtaba 2010). The palaeochannel was confirmed by optically stimulated luminescence (OSL) geochronology showing younger sediments in the low-lying channel part and older aged sediments in upland parts. Figure 4. The channel appears to have been active during ~6-4 ka and might have become ephemeral subsequently (Saini and Mujtaba 2010). Another study indicates the existence of this palaeochannel till 8 ka only (Singh et al. 2017). This palaeochannel can serve as an important zone for researchers and students of Holocene geology and geoarcheology.

Fig. 3 The archeological mounds in Sirsa and Hisar plains are being cut and destroyed by people. The material is used as fillers. Photographed in 2016–17

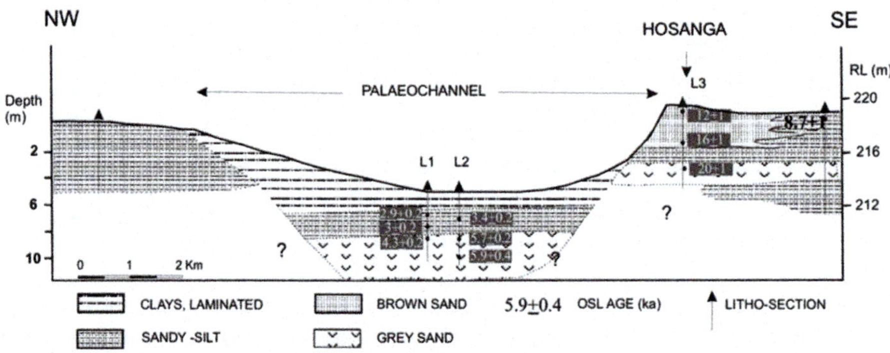

Fig. 4 North-south section across the Sarswati palaeochannel showing a depression which contains fluvial sediments younger than the surrounding upland. (Saini and Mujtaba 2010)

2. **The Dhir Mound**. (N29⁰33'4″; E75⁰28'56″) It is 10 m high prominent mound on the Fatehabad- Ratia road (Fig. 2c) which exposes a 6 meter section on its eastern side (Fig. 5). This section is important because it is the only mound situated in the Saraswati palaeochannel. The section shows two-meter thick natural flood-plain deposit overlain by a four meter thick archeological deposit. The natural deposit contains evidence of human activities in form of soft charcoal,

Fig. 5 Archeological Mound at Dhir, near Fatehabad located in palaeochannel of the Saraswati. It has a substrat of natural deposit overlain by an archeological deposit. The 10,225 yr. BP age is oldest in the region and thus extends the age of habitation in the region. (Alok et al. 2021)

brick pieces and broken bones transported along with silty sediments. Five charcoal pieces transported and deposited in the substrate of Dhir mound at different levels were dated by AMS ^{14}C which gave an age range of cal 10,252 ± 101 to 2030 ± 5 BP carbon year (Alok et al. 2021, Fig. 5) The ages are not in stratigraphic order because the charcoal pieces are transported along with the sediments from the catchment area. A sample of sand at 1.5 m above the base, dated by quartz OSL gave an age of 1.5 ± 0.1 ka (Raza et al. 2020). This mound is important as it preserves, hitherto unreported, presence of ~9 ka old pre-Harrapan anthropogenic activities in the region. The archeological upper part needs to be studied further for evaluating the post Harrapan period in this section.

3. **Mid-Holocene lake Phase archives**: In the desertic and semidesert NW Indian Plains a large number of unique lake deposits are scattered along a 300 km E-W corridor from Hanumangarh district of Rajasthan in west (Karsandi) through Riwasa near Tosham up to Kotla Dahar in Haryana in the east (Saini et al. 2005; Dixit et al. 2014, 2015). These deposits are indicative of a high rainfall period when a large number of small lakes and ponds formed over the aeolian and alluvial plains. In several of these ponds/lakes gypsum and carbonate sediments were precipitated due to high evaporation (Bhatia and Singh 1998). In western arid part the lake deposits are of made up of gypsiferous sediments while in the semiarid eastern part of this tract, these are dominated by carbonate sediments (argillaceous lake limestone, Fig. 6). The gypsum bearing lake deposits suggest a wet period around ~5.1 ± 0.2 ka BP while the carbonate deposits suggest a wet

Fig. 6. A lensoidal lake limestone deposit near Riwasa, Tosham indicative of an extensive early Holocene lake phase. Most of such occurrences have been quarried

phase around ~8.2 kyr BP and another ~5.6 ± 0.3 to 5.3 ± 0.2 ka. ^{14}C yrs. B.P. It was followed by a dry episode at around 4.1 ky BP (Dixit et al. 2014, 2015).

While there is abundance of gypsum lake deposits in Hanumangarh and Ganganagar districts, there are few patches of carbonate lake deposits in Riwasa-Tosham area because most of these have been quarried for a cement factory. These are also encountered in shallow dug wells. There is an urgent need to preserve these as the main geoscientific archives of palaeomonsoon, ecology and palaeogeographical understanding of the Holocene climate fluctuations.

3 Conclusions

Palaeo-rivers and associated features represent primary evidence of past hydrological activities. Combined with evolution and demise of a major civilization, these assume an unparalleled geoheritage significance, The palaeo Saraswati channel, the geological, archeological and composite mounds and the associated evidence of the Harappan civilization require urgent protection which is presently available only to a few archeological sites in the northwestern plains of Haryana and Rajasthan.

These are not only significant archives of Holocene hydrological and climate but also have great tourism potential. Some of the recently identified sites, e.g., the composite geo-archeological section at Dhir, requires special mention in this context.

Acknowledgements Ministry of Earth Sciences is acknowledged for support through a project entitled "A 100 ka glaciation-deglaciation history of Ladakh: its comparison with Sutlej valley and signatures in the alluvial plains of NW India" (MoES/PAMC/H&C/51/2013-PC-II) to NCP, during which part of the work detailed here was carried out.

References

Asrat A, Zwoliński Z (2012) Geoheritage: from geoarchaeology to geotourism. Quaestiones Geographicae 31(1):5–6

Alok A, Pant NC, Das K, Tsutsumi Y, Petrie CA, Kumar P, Chopra S, Saini HS, Khan AA (2021) New insights into the geological evolution of palaeo-rivers and their relationship to the Indus civilization and early historic settlements on the plains of Haryana, NW India. Geological Society, Special Publications, London. https://doi.org/10.1144/SP515-2020-161

Bhatia SB, Singh N (1998) Middle Holocene palaeoclimatic and palaeoenvironmental events in southern Haryana. Proc Indian Natl Sci Acad Part A 54:574–584

Bhadra BK, Gupta AK, Sharma JR (2009) Saraswati Nadi in Haryana and its linkage with the Vedic Saraswati River—integrated study based on satellite images and ground based information. J Geol Soc India 73(2):273–288

Clift PD, Carter A, Giosan L, Durcan J, Duller GA, Macklin MG, Alizai A, Tabrez AR, Danish M, VanLaningham S, Fuller DQ (2012) U-Pb zircon dating evidence for a Pleistocene Sarasvati River and capture of the Yamuna River. Geology 40(3):211–214. https://doi.org/10.18814/epiiugs/2020/020034

Dikshit KN (2012) The rise of Indian civilization: recent archeological evidence from plains of lost river Saraswati and radiometric dates. Bull Deccan College Res Institute 72:1–42

Dixit Y, Hodell DA, Petrie CA (2014) Abrupt weakening of the summer monsoon in Northwest India~ 4100 yr ago. Geology 42(4):339–342. https://doi.org/10.1130/G35236.1

Dixit Y, Hodell DA, Sinha R, Petrie CA (2015) Oxygen isotope analysis of multiple, single ostracod valves as a proxy for combined variability in seasonal temperature and lake water oxygen isotopes. J Palaeolimnol 53:35–45

Giosan L, Clift PD, Macklin MG, Fuller DQ, Constantinescu S, Durcan JA, Stevens T, Duller GA, Tabrez AR, Gangal K, Adhikari R (2012) Fluvial landscapes of the Harappan civilization. Proc Natl Acad Sci 109:1688–1694. https://doi.org/10.1073/pnas.1112743109

Gupta AK, Sharma JR, Sreenivasan G, Srivastava KS (2004) New findings on the course of river Sarasvati. J Indian Soc Remote Sens 32(1):1–24

Hose TA (2018) A wheel along Europe's rivers: geoarchaeological trails for cycling geotourists. Open Geosci 10(1):413–440. https://doi.org/10.1515/geo-2018-0033

Kenoyer JM (1998) Ancient cities of the Indus Valley civilization. Oxford University Press, p 260

Kenoyer JM, Pearsall DM (2008) Indus civilization. Encyclopedia Archaeol 1:715–733

Mahato MK, Jana NC (2021) Exploring the potential for development of Geotourism in Rarh Bengal, eastern India using M-GAM. Int J Geoheri Parks 9(3):313–322. https://doi.org/10.1016/j.ijgeop.2021.05.002

Mehdi SM, Pant NC, Saini HS, Mujtaba SAI, Pande P (2016) Identification of palaeochannel configuration in the Saraswati River basin in parts of Haryana and Rajasthan, India, through digital remote sensing and GIS. Episodes J Int Geosci 39(1):29–38

Nath A (1998) Rakhigarhi: a Harappan metropolis in the Saraswati-Drishadvati divide. Puratattva 28:39–45

Nath A, Law R, Garge T (2014) Initial geologic provenience studies of stone and metal artefacts from Rakhigarhi. Heritage. J Multidiscip Stud Archaeol 2:74–100

Oldham CF (1893) Art. III.—the Saraswatī and the lost river of the Indian Desert. J R Asiat Soc 25(1):49–76

Orengo HA, Petrie CA (2017) Large-scale, multi-temporal remote sensing of palaeo-river networks: a case study from Northwest India and its implications for the Indus civilisation. Remote Sens 9(7):735

Petrie CA, Singh RN, Bates J, Dixit Y, French CA, Hodell DA, Jones PJ, Lancelotti C, Lynam F, Neogi S, Pandey AK (2017) Adaptation to variable environments, resilience to climate change: investigating land, water and settlement in Indus Northwest India. Curr Anthropol 58(1):1–30

Raza MA, Dutta S, Chunchekar RV, Bhavani R, Saini HS, Mujtaba SAI, Hasan SJ (2020) Holocene climate change and its impact on the dispersal of Indus valley/Saraswati civilization. In: Excursion guide book code-NR00436th international geological congress (unpublished), Delhi

Sahai B (1999) Unraveling of the 'lost' Vedic Sarasvati. Memoirs-Geol Soc India:121–142

Saini HS, Alok A, Pant NC (2020) The lost Saraswati River of northwestern Indian Plains: status and way forward. Episodes 43:524–534. https://doi.org/10.18814/epiiugs/2020/020034

Saini H, Mujtaba S (2010) Luminescence dating of the sediments from a buried channel loop in Fatehabad area, Haryana: insight into Vedic Saraswati River and its environment. Geochronometria 37:29–35. https://doi.org/10.2478/v10003-010-0021-5

Saini HS, Tandon SK, Mujtaba SAI, Pant NC (2005) Lake deposits of the northeastern margin of the Thar Desert: Holocene (?) Palaeoclimatic implications. Curr Sci 88:1994–2000

Santangelo N, Valente E (2020) Geoheritage and geotourism resources. Resources 9(7):80

Santangelo N, Amato V, Ascione A, Russo Ermolli E, Valente E (2020) Geotourism as a tool for learning: a geoitinerary in the Cilento, Vallo di Diano and Alburni Geopark (southern Italy). Resources 9(6):67

Sarkar A, Mukherjee AD, Bera MK, Das B, Juyal N, Morthekai P, Deshpande RD, Shinde VS, Rao LS (2016) Oxygen isotope in archaeological bioapatites from India: implications to climate change and decline of bronze age Harappan civilization. Sci Rep 6:1–9. https://doi.org/10.1038/srep26555

Shinde VS, Kim YJ, Woo EJ, Jadhav N, Waghmare P, Yadav Y, Munshi A, Chatterjee M, Panyam A, Hong JH, Oh CS (2018) Archaeological and anthropological studies on the Harappan cemetery of Rakhigarhi, India. PloS One 13. https://doi.org/10.1371/journal.pone.0192299

Singh A, Thomsen KJ, Sinha R, Buylaert JP, Carter A, Mark DF, Mason PJ, Densmore AL, Murray AS, Jain M, Paul D (2017) Counter-intuitive influence of Himalayan river morphodynamics on Indus civilisation urban settlements. Nat Commun 8(1):1617

Sood RK, Sahai B (1983) Hydrographic changes in northwestern India. Man and Environment VII:166–169

Tripathi JK, Bock B, Rajamani V, Eisenhauer A (2004) Is river Ghaggar, Saraswati? Geochemical constraints. Curr Sci:1141–1145

Thakkar A, Jadeja JM, Rajesh SV, Chavan A, Abhayan GS, Bhandari S, Chauhan G, Thakkar MG (2022) Geoarchaeological heritage of Kachchh region, Gujarat, Western India: Geotourism potentials. Geoheritage 14(4):128

UNESCO (2012) The Arouca declaration. Available online: http://www.unesco.org/new/fileadmin/MULTIMEDIA/HQ/SC/pdf/Geopark_Arouca_Declaration_EGN_2012.pdf

Valdiya KS (2013) The river Saraswati was a Himalayan-born river. Curr Sci 104:42–54

Wilhelmy H (1969) Das urstromtal am ostrand der Indusebene und das 'Sarasvati-problem': Zeitschrift für Geomorphologie. Neue Folge, Supplement 8:76–93

Yashpal B (1980) Remote sensing of the lost Saraswati River. Proc Indian Acad Sci (Earth & Planetary Science) 89:317–337

Ladakh Batholith, Trans-Himalaya, India: Subduction Zone Magmatic Products and its Majestic Geoheritage Sites

Santosh Kumar, Sita Bora, and Umesh K. Sharma

Abstract The Ladakh Batholith, as an integral part of the Trans-Himalaya magmatic arc system, is sandwiched between the Indus Suture Zone (ISZ) in the south and the Shyok Suture Zone (SSZ) in the north. The Ladakh Batholith is well known for its subduction zone magmatic records, which consist of variety of igneous rocks viz. gabbro, diorite, granodiorite, granite, aplite and pegmatite intimately associated with volcanics and numerous other geological formations. The subduction of the Neo-Tethyan oceanic lithosphere under the Asian plate during the Cretaceous period resulted in the formation of multiple pulses of calc-alkaline granite magmas that co-existed with mantle-derived mafic magmas. These magmas interacted in deep-seated chambers forming as a series of polychronous, composite plutons making the Ladakh Batholith. This paper outlines briefly the magmatic processes and products responsible for the origin and evolution of the Ladakh Batholith. The Ladakh Batholith is also recognized for its unique geoheritage sites, which may attract many visitors interested in exploring its geological wonders.

Keywords Indus Suture Zone · Shyok Suture Zone · Ladakh Batholith · Calc-alkaline Granites · Mafic Magmas

---

S. Kumar (✉)
Department of Geology, Centre of Advanced Study, Kumaun University, Nainital, India

S. Bora
Department of Geology, Government Degree College, Kanda, Bageshwar, India

U. K. Sharma
Department of Science and Technology, Technology Bhavan, New Mehrauli Road, New Delhi, India

S. C. Tripathi et al. (eds.), *Geoconservation and Geotourism Potential of India*, Society of Earth Scientists Series, https://doi.org/10.1007/978-3-031-81017-6_6

1 Introduction

Granite (*sensu lato*) is a well-known rock occurring in different places across India and is a household name for building stone. However, it results from different geological and crustal evolutionary processes in the Earth's history. It occurs as stocks or plutons or batholiths. The term batholith may consist of single or multiple plutons covering an area of at least 100 sq. km, which may be contiguous or be separated by screens of country rock and commonly named after a prominent geographical location (Clemens and Phillips 2020), used here as Ladakh Batholith (LB). The LB is an integral part of Trans-Himalayan magmatic arc system stretched from Namche Barwa in the east to Nanga Parbat in the west (Fig. 1a), covering an area of about 30,000 sq. km (Fig. 1b). The LB is located between two suture zones-Indus Suture Zone in the north (ISZ) (Fig. 2), Shyok Suture Zone (SSZ) and south of Karakoram Batholith (KB). These suture zones result from the India-Asia collision and subduction of the Indian oceanic plate under the Asian plate. The LB is primarily composed of numerous calc-alkaline multiple intrusions or composite plutons having broad compositional spectrum which range from gabbro-diorite-tonalite to granodiorite-granite-leucogranite-pegmatite (both magnetite and

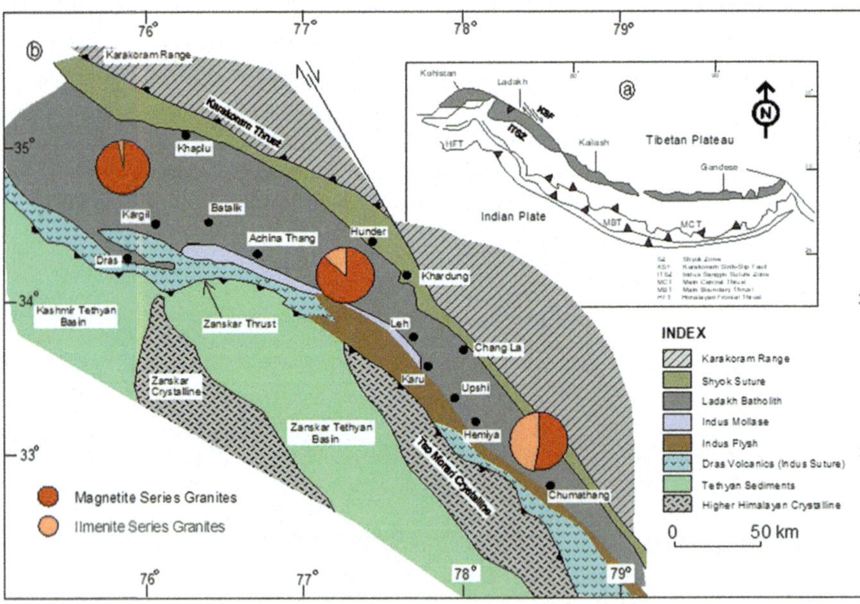

Fig. 1 (**a**) Geological structure of Himalaya showing the location of the Ladakh Batholith as an integral part of Trans-Himalayan magmatic arc system (after Gansser 1977). (**b**) Geological map of Ladakh Himalaya showing the extension of Ladakh Batholith and adjoining lithounits (after Sharma and Choubey 1983; Thakur 1987). Relative abundance of magnetite series and ilmenite series granites in western, central and eastern parts of the Ladakh Batholith is shown by wheel diagrams (Kumar 2008, 2010b)

Fig. 2 Panoramic view of Range of Indus Formation beyond the Indus Valley (Indus Suture Zone), seen from the part of Ladakh Batholith

ilmenite series granites) (Fig. 1b). The LB is cross-cut by several syn- to post-plutonic mafic dykes of basaltic and intermediate (quartz diorite and granodiorite) compositions. The volcanic sequence (basalt-andesite-dacite-rhyolite) overlying the granites of LB is commonly exposed on the way to the Nubra valley from Khardung, referred to as Khardung Volcanics (KV), which records explosive and quiescence nature of volcanic eruptions. Ladakh granites are very well exposed along the Leh-Khardung La road, and its characteristic geological features can be very well studied along this road section. The unique tectonic-geographic location of the LB, coupled with extreme climatic conditions of the terrain, has generated spectacular geoscientific locales for geoheritage and geotourism significance.

2 Tectono-Magmatic Evolutionary Model of Ladakh Batholith

The evolution of LB through time can be very well understood in association with that of the KB. Adequate amount of U-Pb zircon ages exist on the volcano-plutonic rocks of LB and KB and have recently been compiled and re-interpreted along with a new set of geochronological data (Pundir et al. 2020; Kumar and Pundir 2021). The initial subduction of oceanic lithosphere formed the extensive pre-collisional,

Cretaceous or earlier, calc-alkaline granites with the main magmatic pulses during 110–100 Ma (Kumar et al. 2017), forming the parts of the KB in the Nubra Valley along the SSZ. Cretaceous forearc ophiolites are also recognized in the SSZ (Thanh et al. 2012). The LB was formed by the most prominent magmatism, which occurred between 70 and 45 Ma with some discrete early granite pulses at about 100 Ma in the western part of the LB. The KV erupted and deposited mainly during 67–60 Ma, as possible volcanic counterparts of some of the plutonic activity of the LB. The northward continued subduction of oceanic crust and tectonic movement of the Indian plate brought the LB magmatic arc close to the KB at about 85 Ma along the SSZ, and both continued to evolve together before, during and after the collision of Indian-Asian plates. The eastern part of the LB mixed with the syn- to post-collisional granite plutonism forming the profuse leucogranite and pegmatite system.

3 Subduction-Related Calc-Alkaline Magmatism

Mafic bodies of various dimensions are found in close contact with the granites of LB (Fig. 3). At places, these mafic (diorite) masses are disrupted by invading granite-leucogranite-pegmatite system, the outcrops of which are more pronounced in and around the Leh and Sabu regions (Fig. 4). The xenoliths of mafic medium to

Fig. 3 Exposure of granites in the close contact with diorite. In the background Molasses and Indus Formation can be seen that are exposed beyond the Indus Suture Zone

Fig. 4 Leucogranite-pegmatite (light coloured) intrudes the diorite mass (dark coloured)

coarse-grained diorites can be seen hosted within the leucogranite-pegmatite. The southern margin of the Ladakh Batholith is limited by the ISZ having tectonic contact with river-born Indus sediments, molasses, deposited by the ancient Indus River (Fig. 3). The spectacular field features of granite intrusions into the country rocks are apparent towards the northern margin of the LB exposed in the Nubra Valley, where granites intrude the Shyok Group of rocks, mainly into the shale and slate (Fig. 5), indicating the extensive intrusions of different granite magma pulses during and after the closure of SSZ. Profuse leucogranite and pegmatite veins are generated all along the northern margin of the LB exposed in the SSZ (Fig. 6) as a result of the contact of country rocks with the water-saturated granite magma.

The granites of LB are medium to coarse-grained, mesocratic to leucocratic (off-white to steel grey) and equigranular to coarse-grained, and bear common biotite--amphibole-plagioclase-K-feldspar-quartz-zircon-magnetite-apatite mineral assemblage. The proportion of dark-coloured (ferro-magnesian) hydrous minerals (amphibole and biotite) varies from one to another place, and at some places granites are devoid of amphibole. At some places, granites exhibit porphyritic texture with K-feldspar phenocrysts or megacrysts embedded in a medium to coarse-grained mafic-felsic groundmass.

Fig. 5 Ladakh granites, more as leucogranites, intrude the shale and slate of Shyok Formation, exposed in the Nubra Valley

Fig. 6 At and near the northernmost margin of Ladakh Batholith, the intrusion of granites generates profuse leucocratic-pegmatite systems, which host number of xenoliths and blocks of country rocks belonging to the Shyok Formation

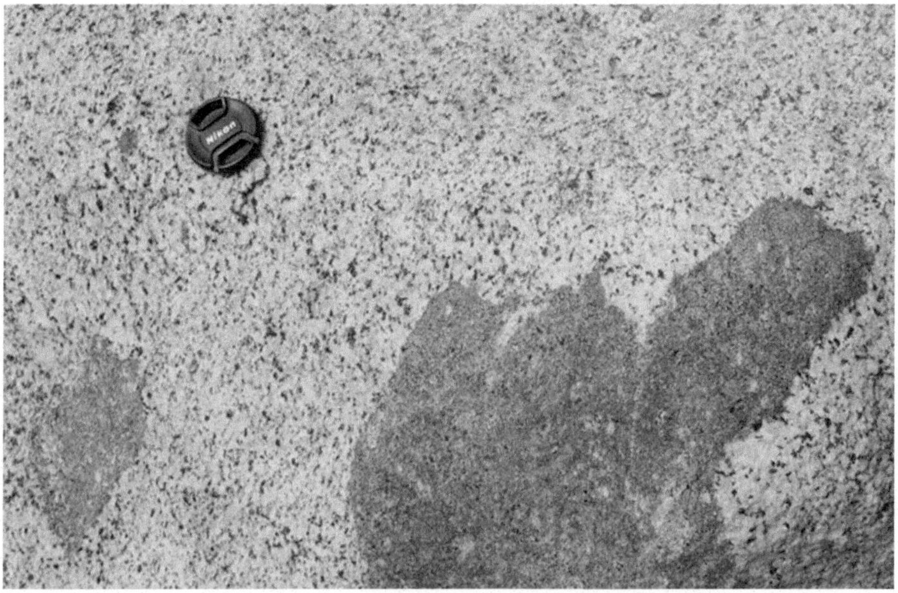

Fig. 7 Granite shows mesocratic colour index nearly equigranular texture with randomly oriented mafic minerals and contains fine grained, occasionally with felsic phenocrysts, melanocratic microgranular enclaves

4 Microgranular Enclaves and Syn-Plutonic Dykes

The most characteristic features of subduction zone magmatism are mafic-felsic magma mixing and mingling in a deep-seated (plutonic) environment that are very well displayed by the presence of ubiquitous microgranular enclaves, mafic to hybrid magma globules of various shapes and sizes hosted in the granites of the LB (Fig. 7). Enclaves refer to any lithic (rock) fragments hosted in granites. These may represent solid rock fragments of lithological units from depth or fragments of country rocks, or aggregates of early crystallized minerals, cumulate or cognate, of host granite magma, or mafic to hybrid (mafic-felsic) magma globules coeval with host granite magma referred as microgranular enclaves. The microgranular enclaves are mingled and interacted with partly crystalline host granite magma at various stages of its evolution within the magma chamber. The enclaves of mafic-felsic magma mixed origin exhibit mesocratic (intermediate between light and dark coloured) to melanocratic (dark coloured) fine-grained to porphyritic textures, rounded to elongate shapes with variable sizes having sharp, crenulate and diffuse contacts with host granites (Fig. 8), which indicate interaction of crystal-charged mafic and felsic magmas at contrasting temperatures and compositions (Kumar 2010a, 2020). Several enclaves cluster at places and follow the direction of magmatic flowage. The disrupted mafic syn-plutonic dyke and sheet-like bodies can also

Fig. 8 Microgranular enclaves of various shapes, sizes and colour indices are embedded in the coarse grained granites, and crudely follow magmatic flowage. Note the sharp to diffuse contacts between the enclaves and host granitoids. Some of enclaves lose its entity and forms mafic schlierens

be recognized (Fig. 9), which are the result of magma injection into the largely crystallized (>65% crystals) of host granite magma.

5 Syn- to Post-Collisional Granite Magmatism

In the eastern Ladakh Trans-Himalaya, certain granitoid bodies of the LB have been caught up within the shear or fault zones. These bodies have undergone various phases of deformation, metamorphism, and melting events during (syn) and after (post) collision of the Indian-Asian plates. The presence of post-collisional magmatic events is evidenced by a network-style of leucocratic (light-coloured) magma movements from depth, which appear as discordant bodies and veins (the hills showing these features popularly called *Zebra Hills*; Fig. 10). These features can be observed in the hills known as Zebra Hills, which suggest syn- to post-deformational melting events of calc-alkaline granites and other lithological units along the Tangtse Shear Zone of Pangong Metamorphic Complex, eastern Ladakh Trans-Himalaya (Fig. 11).

Fig. 9 Syn-plutonic composite mafic dykes and sheets in the granitoids of Ladakh Batholith. These synchronous dykes and sheets are discontinuous and disruptured because of viscosity contrasts between them. The host granite magma achieved high viscosity with increased amount of crystal contents (about >65%)

6 Origin and Evolution of Subduction Zone Magmatism

Granite magmas are commonly generated by partial melting of crustal rocks such as metamorphic, sedimentary and igneous rocks in the Earth's crust under diverse tectonic environments. Igneous rocks cover a wide range of volcano-plutonic rock types. They commonly include gabbro-diorite-granodiorite-syenite-granite and its volcanic equivalent basalt-andesite-dacite-trachyte-rhyolite. The Earth has a continental crust with a differentiated andesite composition, referred to as an andesite planet (Tatsumi et al. 2015). Andesitic magmatism is commonly associated with convergent plate margins such as subduction zones, leading to a consensus that this setting has been the major site of continental crust formation. Continental (felsic) and oceanic (mafic) crust form the bulk of Earth's crust. The upper continental crust is more felsic than the lower continental crust (diorite, amphibolite and tonalite). The melting of these crustal sources produces different types of granite magmas, with characteristics geochemical signatures, in diverse tectonic settings. The magma, thus generated, may subsequently evolve to form a series of igneous rocks with varying silica contents through crystallization and even get contaminated during crystallization with deeper lithological units during its ascent and/or with

Fig. 10 Network of leucogranite-pegmatite veins cross cutting the host granites, amphibolite and occasionally granulites (residuals), which together appear as zebra skin and hence commonly called as Zebra Hills

country-rocks at crustal depth during its emplacement. The granite magma may also experience mixing and mingling with co-existing mafic magma, if any, at various stages of their evolution and formed hybrid rocks in an open magma chamber.

The volcano-plutonic mafic to felsic rocks of Ladakh Trans-Himalaya typically represent arc related calc-alkaline subduction zone magmatism, which are formed by melting of deep subducting Neo-Tethyan oceanic crust and overlying hydrated enriched mantle wedge below the Asian or Eurasian continental plate. The juvenile or young oceanic crust is hot and buoyant (low density) close to its formation site, i.e., mid-ocean ridge and becomes gradually denser with cooling and contraction while moving away from the ridge and may further bend and sink into the underlying mantle along the deep oceanic trench. The composition of the oceanic crust is generally basaltic and contains hydrous mineral amphibole. The dehydration of the subducting slab reduces the melting temperature of the overlying enriched ultramafic rock-bearing mantle wedge that melts partially and produces mafic (basaltic) magma. Substantial partial melting of subducting oceanic crust may occur under high shear stresses maintained by rocks close to, or above, their melting temperatures, which can produce calc-alkaline andesite or adakite-like magmas (Peacock et al. 1994). In most subduction zones, the mafic and felsic magmas thus produced may rise and form the island volcanic arc and/or intrusive igneous rocks, similar to

Fig. 11 Close view of Fig. 10 where discordant (thick light-coloured veins cutting the dark-coloured host rock) and concordant (thin light-coloured veins along the foliation plane of dark-coloured host rock) leucocratic bodies produced by melting of calc-alkaline granites and other lithological units

Khardung Volcanics (basalt, andesite to rhyolite; Figs. 12 and 13) and diorite-granodiorite found abundantly in the LB respectively.

Field, textural and geochemical data narrate a classical evolution of Ladakh volcanic-plutonic complex forming the LB, and postulate that felsic magma emplaced in arc setting. However, there may be genetic links between the volcanic and the plutonic rocks, where volcanics might have erupted as water-saturated differentiated residual products of parental magma, leaving behind fractionated crystal mush rocks hidden at depth as plutonic bodies. The genetic linkages and pre-eruptive histories are most commonly displayed by the phenocrystic assemblage embedded in the glassy or cryptocrystalline groundmass. In the field, the Khardung Volcanics lie over the granites of LB without showing any intrusive or reaction relationships between them. Figure 14 displays the broadly acceptable tectono-magmatic evolutionary model of the Karakoram and Ladakh Arcs in the northwestern Trans-Himalaya, which was proposed by Thanh et al. (2012). The figure illustrates the processes that have shaped these arcs over time and provides a detailed account of their geological history, as explained in the preceding paragraphs.

Fig. 12 Flow layers of Khardung volcanics can be recognized in the field, mostly shown by basalt, basaltic andesite and dacite lying above the LB on the way to Nubra from Khardung

7 Geotourism Potential of Ladakh Batholith

The unique combination of geographic setting, tectonic milieu and climatic conditions provides some spectacular sites associated with the above-described geological and petrological context. A few of these are detailed below.

Geoheritage Site 1: Intrusion of granites (Ladakh Batholith) into shale/slate (Shyok Formation).

Location: Hunder (N 34°32′48″, E 77°33′36.3″) located near iron bridge and Gompa on the way to Turtuk in the Nubra valley.

Significance of site: leucocratic granites intrude the metasedimentary (metarenaceous and metapelites viz. shale and slate) country-rocks as exhibited by the presence of disrupted xenoliths of shale/slate with irregular outlines (Fig. 5) along the northern margins of the Ladakh Batholith. At this place leucogranite converts to ilmenite (reduced) series because of its marginal reaction with pelitic rocks. Profuse pegmatite and aplite veins occur near the intrusive contacts of the Ladakh granites with the Shyok country rocks because of devolatization as soon as it comes into contact of cooler country rocks. U-Pb zircon crystallization age (67.32 ± 0.66 Ma) of leucogranite implies that the Shyok metasediments (shale/slate) were deposited and metamorphosed prior to ca. 68 Ma.

Geoheritage Site 2: Mafic to hybrid synplutonic dykes hosted in the granite illustrate the features of granite magma chamber intermittently recharged with mafic to hybrid magmas when the host granite magma crystallized more than 65%.

Fig. 13 Rhyolite volcanism exhibits reddening due to strong oxidation. The background hills belong to Karakoram Block that is uplifted more than the Ladakh Block

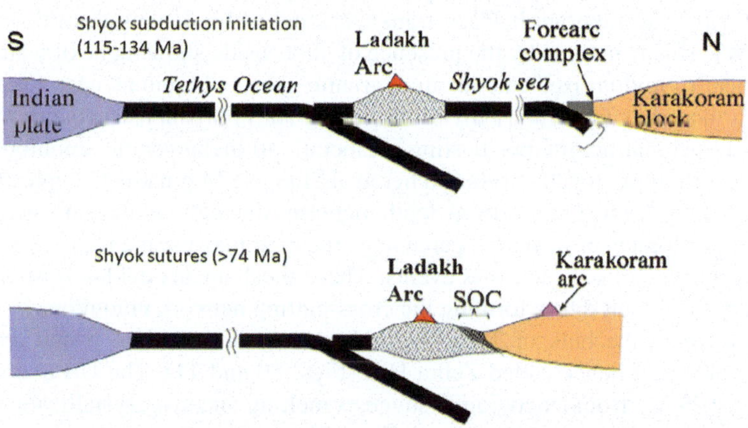

Fig. 14 Tectono-magmatic evolutionary model for the northwestern Trans-Himalaya through time forming the Karakoram and Ladakh Arcs (after Thanh et al. 2012). (**a**) Northward-dipping subduction zones beneath the southern Ladakh–Kohistan arc and the Karakoram block in the Early Cretaceous. (**b**) The collision of the Ladakh Arc and the Karakoram Block in the middle to late Cretaceous

Location: Taru (N 34°11′19.7″, E 77°23′59.9″) located at the end of cantonment area on the northern side of Leh-Dras national highway (NH1D), close to the Leh Campus of University of Ladakh.

Significance of site: Magma chamber processes at shallow depth i.e., ca 1 km (75–135 MPa):

Mafic and hybrid magmas intrude the crystallizing host granite magma, having more than 65 vol. % crystals. Because of compositional and rheological contrast, the intruding mafic and hybrid magmas got disrupted to form composite platy, sheet-like synplutonic dykes and enclave swarms, which are broadly oriented together in a dyke strike direction (Fig. 9). These are undercooled, forming a pillow-like margin against relatively cooler host granite. The synplutonic hybrid dykes contain felsic xenocrysts that are sourced and injected from a hybrid magma zone hidden at depth. The conventional geological knowledge is that the intruding dyke should be younger than the host granite. However, the determined U-Pb zircon ages from mafic (57.61 ± 0.67 Ma), hybrid (57.08 ± 0.45 Ma) synplutonic dykes, and host granite (57.19 ± 0.38 Ma) are the same within the given associated errors that strongly underline their coeval nature.

Geoheritage Site 3: Zebra Hill showing the network style of leucogranite magma ascent from 15 km depth to higher crustal level, and in situ melting.

Location: Tangste (N 34°01′34.6″, E 78°12′07.7″) to Pangong road in the Tangste Shear Zone.

Significance of site: Tangste Shear Zone (TSZ) separates the Pangong Metamorphic Complex (PMC) to the northeast from the Ladakh Batholith to the southwest. The TSZ is an integral part of Karakoram Shear Zone that lies in the central part of Karakoram Fault of trans-tensional nature having dextral strike slip (N55° W). This is marked by the presence of slickenside showing dextral strike slip movement of mylonitized calc-alkaline granite gneiss ubiquitous near Tangste village. Farther away from this, a mylonitic zone grades into intense network of leucogranite to pegmatitic granites hosting mesocratic to melanocratic amphibolite and granulite lithotypes together referred herein as Tangtse Migmatite Complex (TMC). Melting of protolith at mid-crustal depth increases the melt volume and after attainment of rheological critical melt percentage the melt may escape from the source in batches during several melt-loss events. These mechanisms evidently support the occurrences of melt dykes forming the cross-cutting network ubiquitous in Tangtse region forming the bulk of TMC in the higher crustal section, which appear as Zebra's skin and hence called Zebra Hill (Figs. 10 and 11). The U-Pb zircon age 18.27 ± 0.29 Ma from leucogranite suggests melting and syn-crystallization events during the development of Karakoram Shear Zone.

Acknowledgments The present work is the outcome of the projects on the DST field training programme *(SR/FTP/ES-141/2010)*, SERB research grant (EMR/2017/002420) to SK, and SERB N-PDF grant *(PDF/2021/002802)* to SB. The generous comments from N. C. Pant significantly improved the earlier version of the manuscript. We thank the Series Editor Satish C. Tripathi and volume Editor N C Pant for inviting this contribution. The DST field training programme resource persons and participants generated insightful and valuable geoscientific discussions.

References

Clemens JD, Phillips GN (2020) Granite suites: a problematic concept? Aust J Earth Sci 67:509–523. https://doi.org/10.1080/08120099.2020.1701076

Gansser A (1977) The great suture zone between Himalaya and Tibet: a preliminary account. Sci Terre Himalaya CNRS 268:181–192

Kumar S (2008) Magnetic susceptibility mapping of Ladakh granitoids, northwest higher Himalaya: implication to redox series of felsic magmatism in the subduction environments. Memoir Geol Soc India No.72:83–102

Kumar S (2010a) Mafic to hybrid microgranular enclaves in the granitoids of Ladakh batholith, northwest higher Himalaya: implications on calc-alkaline magma chamber processes. J Geol Soc India 76:5–25

Kumar S (2010b) Magnetite and ilmenite series of granitoids of Ladakh batholith, northwest Indian Himalaya: implications on redox conditions of calc-alkaline felsic magma chambers. Curr Sci 99:1260–1264

Kumar S (2020) Schedule of mafic to hybrid magma injections into crystallizing felsic magma chambers and resultant geometry of enclaves in granites: new field and petrographic observations from Ladakh batholith, trans-Himalaya, India. Front Earth Sci 8:551097. https://doi.org/10.3389/feart.2020.551097

Kumar S, Pundir S (2021) Tectono-magmatic evolution of granitoids in the Himalaya and trans-Himalaya. Himal Geol 42:213–246

Kumar S, Bora S, Sharma UK, Yi K, Kim N (2017) Early Cretaceous subvolcanic calc-alkaline granitoid magmatism in the Nubra-Shyok valley of the Shyok Suture Zone, Ladakh Himalaya, India: Evidence from geochemistry and U–Pb SHRIMP zircon geochronology. Lithos 277:33–50

Peacock SM, Rushmer T, Thompson AB (1994) Partial melting of subducting oceanic crust. Earth Planet Sci Lett 121:227–244

Pundir S, Adlakha V, Kumar S, Singhal S (2020) Closure of India–Asia collision margin along the Shyok Suture Zone in the eastern Karakoram: new geochemical and zircon U–Pb geochronological observations. Geol Mag 157(9):1451–1472. https://doi.org/10.1017/S0016756819001547

Sharma KK, Choubey VM (1983) Petrology, geochemistry, and geochronology of the southern margin of the Ladakh batholith between Upshi and Chumathang. In: Thakur VC, Sharma KK (eds) Geology of Indus suture zone of Ladakh, 204p, Wadia Institute of Himalayan Geology, Deharadun (India), pp 41–60

Tatsumi Y, Sato T, Kodaira S (2015) Evolution of the earth as an andesite planet: water, plate tectonics, and delamination of anti-continent. Earth Planets Space 67:67–91

Thakur VC (1987) Plate tectonic interpretation of the western Himalaya. Tectonophysics 134:91–102

Thanh NX, Rajesh VJ, Itaya T, Windley B, Kwon S, Park C-S (2012) A cretaceous forearc ophiolite in the Shyok suture zone, Ladakh, NW India: implications for the tectonic evolution of the Northwest Himalaya. Lithos 155:81–93

A Sustainable Approach Towards Preservation of Geoheritage and Geodiversity: A Case Study of Patiya Village Cluster in Almora District of Uttarakhand

Prabhas Kumar Pande, Rakesh Kumar, Seema Gupta, Rajesh Sachdev, Mukta Datta Mazumder, Kalpana Gururani, Upendra Vikram Maurya, and Naresh Chandra Pant

Abstract In the present context of Geoheritage being under constant threat due to absence of development planning guidelines at the global level, geotourism has the potential not only to play a crucial role in the conservation of geoheritage but also sustainable development of any area of significant geological value through a regulated tourism activity. Cultural, social and economic aspects are important considerations, which can make geotourism a powerful mode of sustainable tourism particularly in the rural areas of Himalaya. Globally, recreational aspect has been dominant in the geotourism sites developed so far, which are very exclusive in displaying some rare landform or process of nature. Geotourism potential of places with pure geo-educational values supported by the local community has not been tested yet. A Geo-excursion site developed in the Patiya village cluster in Almora district of Uttarakhand is not only significant from a cultural and social point of view but also has great potential to attract young geotourists to pursue field work on different themes of geological studies in the Kumaon himalayan region. The present article highlights the different phases of the development of a geo-site involving identification of geological traverses, development of affordable accommodation

P. K. Pande (✉)
Department of Geology, Ram Lal Anand College, University of Delhi, New Delhi, India
e-mail: prabhaspandey.geo@rla.du.ac.in

R. Kumar
Department of History, Ram Lal Anand College, University of Delhi, New Delhi, India

S. Gupta · R. Sachdev · M. D. Mazumder
Department of Statistics, Ram Lal Anand College, University of Delhi, New Delhi, India

K. Gururani · U. V. Maurya
Ram Lal Anand College, University of Delhi, New Delhi, India

N. C. Pant
Centre for Advanced Studies in Geology, University of Delhi, Delhi, India

S. C. Tripathi et al. (eds.), *Geoconservation and Geotourism Potential of India*, Society of Earth Scientists Series, https://doi.org/10.1007/978-3-031-81017-6_7

(home stays) and community participation. It showcases the efforts that have gone into developing a sustainable geotourism model in a remote Himalayan village cluster that can also be replicated over many parts of Himalaya having a great geodiversity that needs to be conserved.

Keywords Geo-excursion · Sustainable geotourism · Almora · Home stays

1 Introduction

Since centuries, admiration of scenic and spectacular landforms and geological process (geodiversity) along with associated diverse life (biodiversity) has motivated tourists to explore such places. This great connect between humans and nature has resulted into ever-growing tourism industry which plays a key role in the economy and development of a region though often with adverse impact on environment due to lack of appreciation of sustainability component in these ventures. Last few decades have witnessed a great amount of research to understand the intrinsic relationship between geodiversity and tourism, which not only reinforces the definition of Geotourism as a sustainable form of tourism, but has also, activated the Earth scientists to contribute significantly on importance and conservation of Geoheritage world over.

2 Geo-tourism, -diversity, -heritage, -conservation: An Inherent Relationship

Geotourism is a relatively new form of niche tourism. Its definition has undergone many modifications since its inception (Hose 2005, 2016; Larwood and Prosser 1998; Dowling and Newsome 2010; Dowling 2013). Early definitions were intricately linked to visit and interpretation of geo-sites with recreational value, but conservation and interpretations of abiotic components were also added to it later (Nekouie-Sadry 2009). Hose (2016) presented "4Gs model of Geotourism" which includes Geoconservation, Geohistory, Geointerpretation and Geosites/geomorphosites as its main components making it more comprehensive. However, it is likely to get modified from place to place due to the impact of prevailing forms of tourism, for example, in India, religious tourism is the most prevalent form of tourism and its proximity with nature can only be understood with historical development of the society (Pande et al. 2018).

Earlier, the concept of Geodiversity was focused on major abiotic component of nature helping biodiversity to develop on it. Later its definition was broadened based on its unique relation with one or another aspect of nature (Stanley 2000, 2004; Burek 2001; Prosser 2002; Gray 2003, Burek and Prosser 2008). In a way,

Geodiversity refers to all geological material, landforms and natural processes possessing scientific, ecological, aesthetic, cultural and economic values (Bâca 2015).

Geoheritage and Geoconservation are two inherent values embedded in the concepts of Geodiversity and Geotourism which strongly advocate protection of significant Earth history (Sharples 2002; Brilha 2016; Gray 2013). The International Union for Conservation of Nature (IUCN) is the leading global authority on the status of the natural world and the measures needed to protect it (http://www.iucn.org/about). The interest of IUCN in Geo-conservation has increased in the last decade as revealed by the approval of three Geo-heritage-focused resolutions in 2008, 2012 and 2016. In 2014, Geoheritage Specialist Group under the IUCN World Commission on Protected Areas (IUCN-CPA) was formed to provide assistance on all aspects of Geodiversity in protected areas and this group is also involved in developing the "best practice guidelines on Geoheritage conservation" (https://www.iucn.org/commissions/world-commission-protected-areas/our-work/).

The incorporation of a chapter on Geodiversity and Geoconservation in the "Protected Area Governance and Management" handbook (Bertzky et al. 2015) can be seen as consistency on part of IUCN, which is also responsible for the assessment of the new entries with "Outstanding Universal Value" of geoheritage for inclusion in UNESCO's World Heritage sites, together with a strong emphasis on cooperation regarding UNESCO Global Geoparks. The efforts made world over to emphasize the potential of Geoparks and sustainable tourism development plan of many places (mostly in European countries) were significant to draw the attention of government agencies and people to a new impactful level (Frey 2003; Andrasanu 2005; Alexandrowicz 2006; Brilha 2009; Corte 2008; Cottrell et al. 2007; Eder 2004; Ielenicz 2009; Joyce 2010; Farsani et al. 2011; Hose and Vasiljević 2012; McKeever and Narbonne 2021). It can also be seen as an impressive initial attempt to develop a desired perception in society on Geoheritage, which has a considerable connect between tourism, research and education (Andrasanu 2005; Corte 2008; Eder 2008; Fassoulas 2003).

Preparation of an inventory of geodiversity and geoconservation has strong traditions in some developed countries, which implement well drawn Geodiversity Audits and Action Plans for the conservation of the various geosites. India is an exception where, despite the presence of numerous potential geoheritage (including many sequences with stratigraphic value) and geotourism sites, no such conservation policies exist. Scientific and educational values are essential components of Geodiversity (Gray 2004; Brocx and Semeniuk 2007; Brocx et al. 2019) and it has been argued that the greatest threat to geodiversity is ignorance at different levels of society.

In Indian society, little sense of realization about our geological heritage prevails outside academia and the Geological Survey of India (GSI). In India, the efforts made by GSI in declaring some of the Geosites as National geological monuments have somewhat been able to spread awareness in society, but this status has been conferred only to a few rare geomorphological, lithological, and fossil sites (https://www.gsi.gov.in/.../pageGEOTOURISM). All such locations in India are looking for a self-sustaining model to meet the challenges, in the absence of which it becomes a non-productive endeavour. Moreover, despite the presence of numerous geoheritage sites, no part of Himalaya, unfortunately, falls in this list, perhaps due to inaccessibility and lack of infrastructure. Efforts made by INTACH, through a monograph, are commendable where the need to develop and conserve these geological identities as a source of scientific knowledge has also been advocated (INTACH 2016). Many academicians and researchers have significantly highlighted the scientific and tourism potential of many places throughout India (Ahluwalia 2006; Mazumdar 2007; Swarna et al. 2013; Kale 2015; Shekhar et al. 2019). However, so far, the tourism industry and administration has largely failed to realize its potential as a lucrative venture as well as a mechanism to generate additional livelihood options in remote parts of the country. In the absence of any legislation for Geo-conservation and viable Geo-tourism model, these efforts have not produced any significant positive outcome.

Geology is a field-based science, but implementation of its curriculum has been lackluster in many Indian institutions of higher education on account of complex factors, such as poor preservation, obliteration, encroachment of well recognized geoeducation locales etc. A sustainable solution can be to develop an integrated geoeducation cum geotourists locale with the active involvement of local stakeholders having socio-economic components embedded into it.

3 Geo-excursion Site: Scope and Definition

(a) **Scope of Geo-excursion site**

A geo-excursion site provides an opportunity for students to understand rocks in their natural settings. Universities and Institutes offering Geology as a program at the undergraduate and postgraduate level in India are facing an ever-increasing problem due to rapid and often unorganized urbanization, which has resulted in fast depletion of geological heritage sites. Developmental considerations such as expansion of mining sector, development of townships to meet the infrastructural demands of growing population etc. have impinged upon the rock exposures, which have been places of learning for geology students throughout India. Some of the prominent examples of geo-sites in India are Alkaline rocks (Nepheline Syenite) of Kishangarh, Rajasthan; Charnockite queries of south India; Permian-Triassic transition sequence in Guryrul ravine, Srinagar, Kashmir; Mesozoic/Cenozoic sequences of Kutch region etc. In the absence of an effective national policy to protect these places, increasing necessity is being felt for sustainable development of geologically important places required for basic field training. A desired site requires provision for a safe and comfortable stay, along with presence of interesting and educative or unique geological material in natural settings for learning the various theme based geological field aspects. This unique learning site needs to be managed in a sustainable manner by the local people who would also act as custodians of the site.

(b) **Defining the Geo-excursion Site**

An ideal geo-excursion site should be one where several geological features can be witnessed within a small area, which is safe and accessible for field activities and is surrounded by natural beauty with meadows and water springs. In addition, a geo-excursionist would need a decent place for food and accommodation located near the site. The impact of a developed Geo-excursion site would be three dimensional—Preservation of Geo-heritage, Education and Research and Socio-economic development of the local community.

(c) **The Sustainability Factors in Geo-tourism: Economic, Cultural and Educational Dimensions**

1. **Rural Community**

Touristic activities, especially in the hill areas of India and abroad has increased over the years. In India, while economic benefits of tourism have been largely harvested by the urban populace in the hills, the same has not percolated down to the rural community. Speaking specifically of Uttarakhand, the mountain villages, in the past few decades, have seen increasing number of out-migration due to limited employment opportunities. The existing villagers are surviving on subsistence-based agriculture. The agricultural activity in this region is sluggish as the yield is low and uncertain due to several factors that include climate change induced erratic rains, unscientific methods of farming and human-wildlife conflict. To arrest out-migration and improve standard of living of the residents of mountain villages, there is an urgent need to introduce alternate and sustainable employment opportunities.

Geo-tourism, very less explored in India so far, can prove to be an effective game changer for the rural community. Since most of the geo-excursion/geo sites are situated in the rural areas with limited infrastructural facilities, the geo-tourist would primarily require a place for accommodation and food. Several heritage houses (some more than 100 years old) with beautiful traditional architecture in the region are either lying vacant or underutilized. Such houses can be converted into home stays or affordable accommodation for the tourists thus adding to the income of the house owners. Besides, requirement of food, trekking guide and transport services would generate additional employment for the villagers.

In addition, the villagers can use the stay period of the geo-tourists to display their hospitality and showcase local culture as well as sell their handicraft products thus creating additional customers for the local produce alongside preserving their cultural heritage, both tangible and intangible. As geo-tourism in the locality takes its roots, the cooperative efforts would help sustain this activity and economic benefits would propel them to preserve the Geoheritagein the locality. Thus, in the absence of any legislation to protect the rocks, the rural community would prove to be an effective interest group in the preservation of geo-diversity of the region. The one hidden positive outcome of such activity will also be the preservation of local history and culture.

2. **Coupling of Geo-tourism and Geo-education**

   Since, field study is an integral and essential part of the curriculum for the students and researchers of Geology, the demand for places ideal for field study would be never ending. As Geo-tourism is knowledge-based nature tourism, the geo-excursionists/tourist in this case would be presumed to be educated and sensitive about geo-diversity of the region.

   The geotourists, for the sake of convenience and economic considerations, would prefer to stay near the geo-site where affordable accommodation with modern facilities is available. Location of several geo-sites in proximity of each other would obviate the necessity of transport once the tourist reaches the locality. An affordable accommodation with food would motivate the tourists to conduct field study and research for a longer duration. In addition to the geo-sites, the tourists can be given a guided tour of the nearby heritage structures (ancient temples, water reservoirs etc.) which would give them a measure of the history of the locality. Besides, the tourist would get an opportunity to interact with local inhabitants and learn about the local culture and cuisine and enjoy their hospitality in addition to the pure air that they would enjoy against the polluted city environment that they may have come from. At the same time, the tourist would also get an opportunity to purchase organic local products at reasonable rates to take home.

   Since such type of tourism would, in the long run, assist in the growth of rural tourism and lead to rural development, therefore, besides trainings imparted to the home stay owners and other hospitality service providers, the village panchayat, teachers, youth, children as well as women would also need to be trained as a part of the broader stake holder group. On the other hand, the geo-tourists would also be expected to follow responsible tourism.

(d) **Preservative Aspects of Geo-tourism in Geo-excursion site**

The notion of nature conservation has invariably been linked with preservation of its biodiversity (living things). It is only recently that earth scientists across the globe have started recognizing the importance of geo-diversity (landforms, soil, and geological features) and its direct impact on the biodiversity. This has made both, scholars, and scientists, to consider geo conservation as an integral part of bio-conservation. Preservation of geo-heritage, therefore, becomes a crucial and imperative step in the nature conservation.

In this direction, the major thrust area in India should be preservation of Geoheritage in the Himalayas. The hills have always attracted tourists from across India and abroad with attendant benefits and harmful consequences in such regions. The Himalayas (particularly in the Uttarakhand region of India) fall in the high seismic zone which makes the entire region ecologically sensitive. The unplanned tourism infrastructural developments and indiscriminate mining of rocks has further threatened the already fragile Himalayas. The climate change in the form of high precipitation during a short span of time has further compounded the problem. The consequences (landslides etc.) of bad planning, public and private mining of rocks, and climate change have been all too evident lately in the Uttarakhand region in the recent few years. While tourism in the hills cannot be halted for its economic significance, there is an urgent need to develop new forms of sustainable tourism in the mountains.

Geo-tourism, with its eco-sensitive nature is certainly a viable model of sustainable tourism. Sustainability is embedded in the very definition of geo-tourism propounded by scholars and earth scientists. It primarily aims at conservation of Geoheritage besides its educational advantages for students and researchers alike. Since this form of tourism comes under the category of niche tourism, only genuinely interested tourists (students and researchers for their field study) would visit the geo sites thus preventing harmful effects of overcrowding of tourists. The economic benefits (discussed below) accruing from this type of tourism will form the basis of its sustainability as well as ensure community preservation of Geoheritage in the region. Extended further, the model would go a long way in preservation of Geoheritage at pan India level.

4 Site Area Under Development: Description, Methodology and Progress Made

Realizing the need of field work in Geo-education in India, authors were motivated to develop a geo-excursion site in one of the remote village clusters in Lesser Himalaya under a project funded by National Mission on Himalayan Studies in 2019 to provide a new dimension to Geoeducation, Geoconservation and Geotourism involving sustainable aspects mandatory in the Himalayan terrain.

The sites, located in Lesser Himalaya is approached by passing through HFT and MBT with several locations on the way to observe important structural and geomorphic features to understand the formation, sedimentation and tectonics of Himalaya (Célérier et al. 2009). Besides, several igneous intrusions and metamorphic rocks make it more interesting to learn these in consonance with Himalayan tectonics. In addition, the nearby regions of geological importance include exposures of Almora Granites, North Almora Thrust (NAT), South Almora Thrust (SAT), Main Boundary Thrust (MBT) etc.

Geoexcursion site of the present study is situated about 12 km. NE of Almora on SH6 (NH309) leading to Bageshwar via Takula. The focus area of the project i.e. Patiya Village Cluster, falls in the physiographic-cum-geological unit of the Lesser Himalaya and provides an excellent opportunity to develop it into a unique learning site to conduct geological field studies for students of Geology from all over India and abroad as well as a site for discerning nature loving tourists.

A road from Pilkha to Kasoon village (off Sundarpur on Almora-Binsar Road) has exposed the fresh rock outcrops revealing several aspects of Earth history

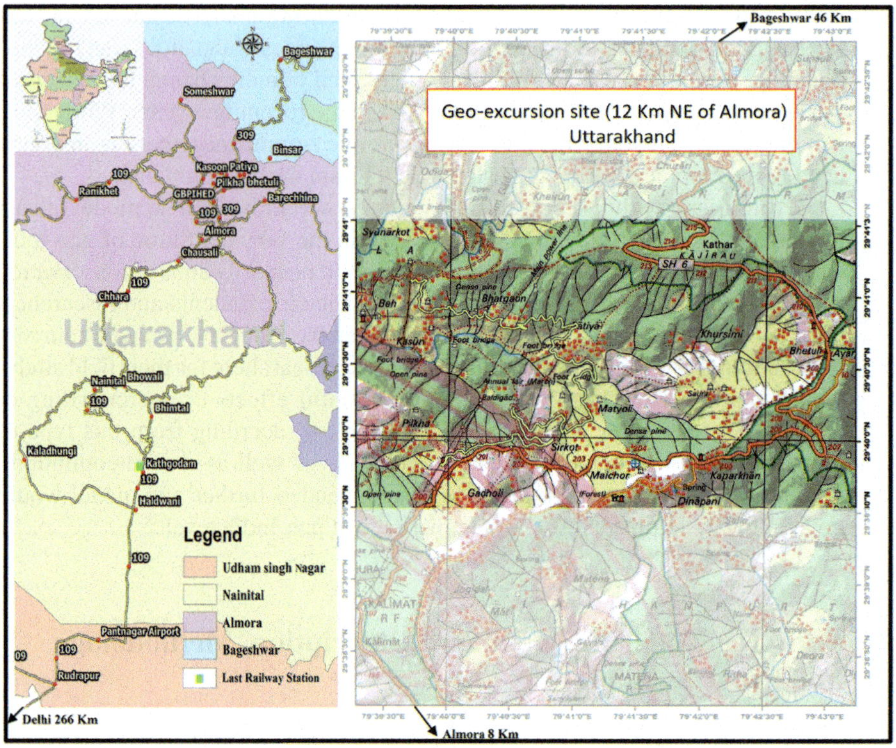

Fig. 1 Location and direction map of the Geo-excursion site developed in Almora district of Uttarakhand. Image on the right shows the location of Geo-excursion site in topo-sheet (Yellow road depicts the main traverse in the site)

(Figs. 1 and 2). The site contains several rock exposures for basic field training, geological mapping, structural geology, geo-morphological studies within a travelling distance of 4–6 km. Hairpin bends along the road have exposed many large outcrops with prominent inclined bedding planes making it an ideal site to understand the basic concepts of bedding plane, the attitude of beds and their measurement (dip-strike and thickness). Presence of several sets of joints in rocks, series of folds in a small part of site and a local thrust passing by, gives a unique opportunity to learn many aspects of Structural Geology like Fold geometry, types of folds, identification and measurements of Primary (scalars and vectors) and secondary structures (linear and planar) and their trend, plunge, Rake/Pitch etc. (Fig. 3). In addition, presence of structural deformations provides an opportunity to identify fold types and their phases.

Identification of rock types and their characteristics; depositional environment; stratigraphic position and understanding of regional and local geologic structures formed because of large- and small-scale deformations, gives an opportunity to know the structural evolution of the region.

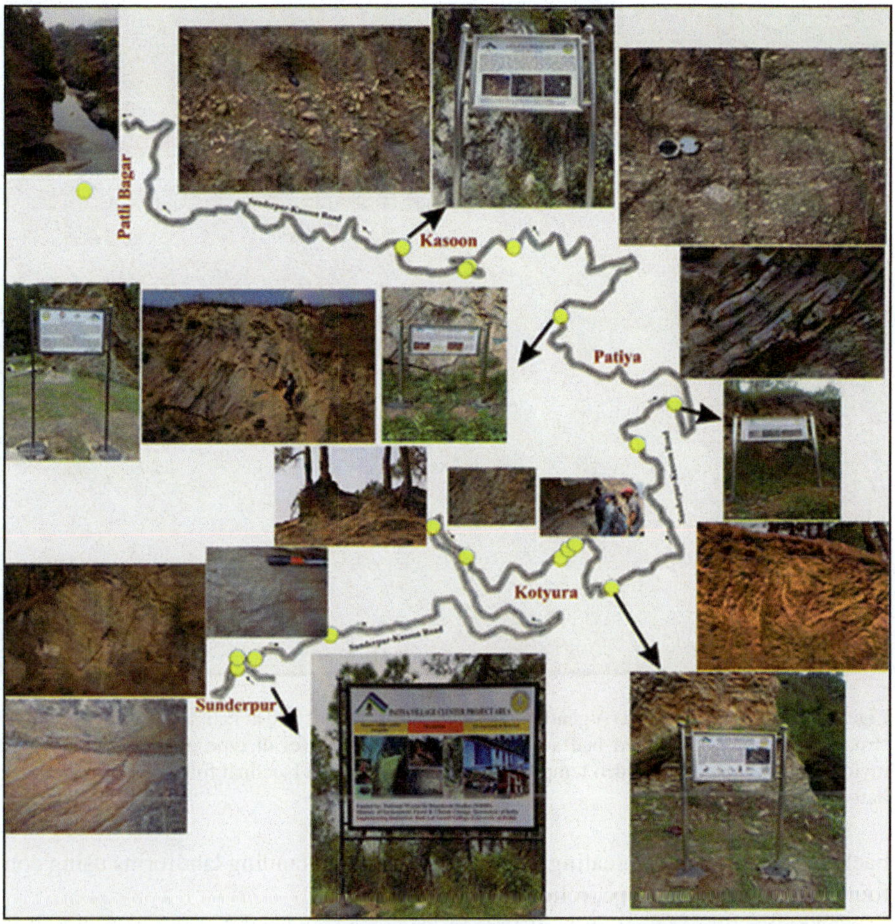

Fig. 2 Overview of the site indicating identified Geological traverse with exposure locations (marked in yellow circles), geo-information boards, rock protection boards and some rock exposures (at Patlibagar, Kasoon, Patiya, Kotyura and Suderpur) to highlight the conceptualization of Geo-excursion site at Patiya village cluster, Almora

The view of the Kumaon Himalayas with varying relief and drainage networks provides good observation of geomorphic features. The Patiya River and the adjoining areas form an identifiable fluvial system (Fig. 4). The site also gives an opportunity for identification and mapping of various geomorphic features like valleys, rivers, and in-stream geomorphic features. Various measuring parameter related to catchment like morphometries, longitude profiles, river cross section, terraces and depositional analysis, drainage discharge etc. can also be performed.

The exposures allow for preparation of lithologs and detailed mapping of the region. This, combined with available topographic maps can be used for gaining marginal information along with toposheet exercises like locating position with

Fig. 3 Exposures showing (**a**) A gently plunging large synclinal fold at Patiya, Almora (**b**) Mullion structures formed in inclined bed section, Kotyura, Almora (**c**) Φ type grain tail complex in mylonite exposure near Samdeo temple, Bhatgaon, Almora (**d**) Isoclinal fold near Patiya Bridge, Patiya, Almora

back bearing exercises, creating cross sections, understanding landforms using contours, their structure, slope, relief, drainage etc.

The major lithological units of Precambrian rocks exposed in this region belong to the Sarju and Gangolihat Formation near Patiya and Bhatgaon Village respectively (Puniya et al. 2019). The rocks of the Sarju formation belong to the crystalline rocks of the Almora Group and are constituted of mica-schist, garnetiferous mica-schist, chlorite sericite schist, massive slaty-quartzite, friable quartzite and slate. The rocks of the Gangolihat Formation are made-up of dolomite, dolomitic limestone and limestone with bands and intercalations of blue limestone and gray slate.

Basic field training for a geo-excursionist will revolve around use of toposheet, Brunton compass and GPS to learn contour patterns on toposheet, orientation of the toposheet in field and marking locations on it with the help Brunton/ GPS. Identification of rock types, bedding and structures and measurement of dip and strike will provide students an ideal learning platform in natural settings.

Fig. 4 A collage of five images showing geomorphological features. (**a**) Various fluvial geomorphic units and features at Guldar valley, Patlibagar, Almora (**b**) Ribbon type waterfall near inclined bed section, Kotyura, Almora (**c**) Potholes in Patiya River at Kasoon, Almora (**d**) Point bar in Kosi River, Guldar valley, Patlibagar, Almora (**e**) River Patiya displaying Bed Rock River features

Presence of a Magnesite mine about 36 km from the site, gives an additional opportunity to study mineral deposit and their formation process in an opencast mine along with a chance to see stromatolites, dolomites and few more lithologies and associated sedimentary structures present in the mine area. Beneficiation plant located here would add to the hands-on knowledge of mineral processing and products marketed by the Almora Magnesite Limited.

Presence of rich Geo-diversity in this Geo-excursion Site makes it a unique place, in terms of providing a range of field-oriented learning aspects of Geology (Fig. 5), which will certainly become the favourite destination of many institutes of India in the near future.

The area is not only significant for geo-education related tourism but also has sites of interest for nature loving tourists as it has spectacular geomorphic sites of mountain river development, major tectonic structure of the Himalayan mountain belt and also provides picturesque view of well-known Himalayan peaks from the newly developed viewpoint within the site with a binocular facility.

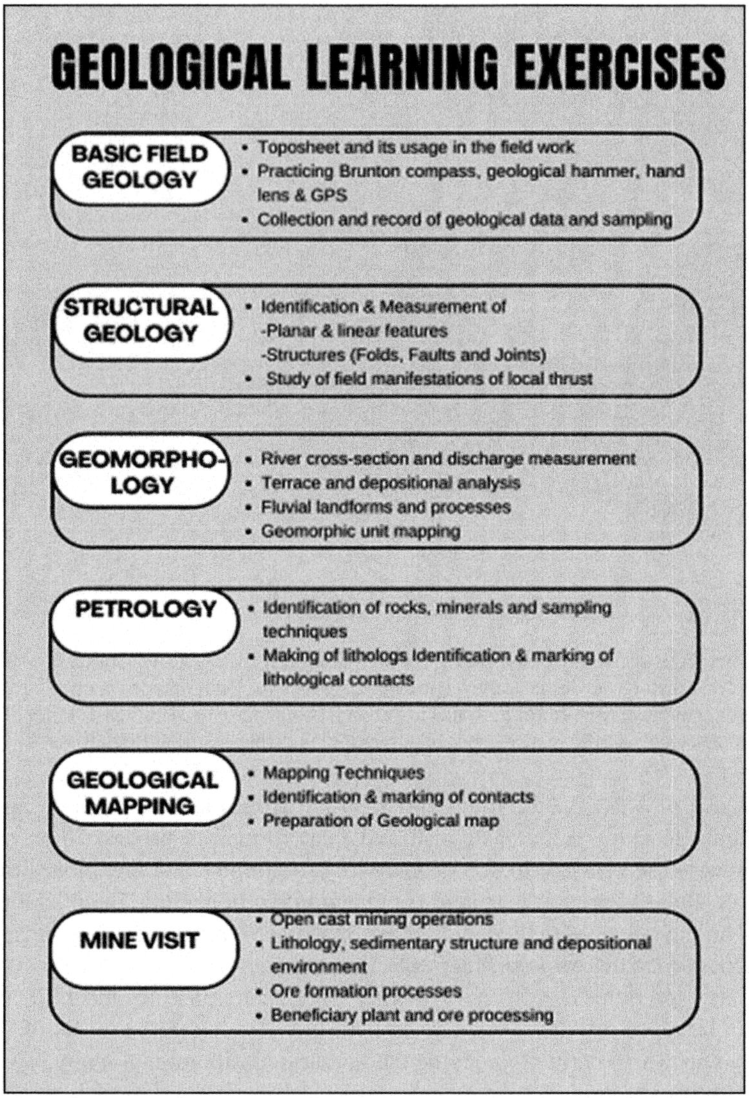

Fig. 5 Depiction of educational aspects of the Geo-excursion site

5 Methodology

A three-phase methodology has been adopted for the project that include—(i) Identification of the site, analysis of its geo-tourism potential and its development as geo-excursion site, (ii) Development of home stays as affordable accommodation and as a source of additional employment, and (iii) Involvement of the local community.

First phase: An extensive study of literature related to tourism and its ecological impact; Geodiversity; Geo-conservation and Geotourism; laws and policies for the protection of rocks and Geoheritage in the world and in India was undertaken.

- The literature review was followed by regular field surveys to identify easily accessible and picturesque sites with snow-clad Mountains in the backdrop for an offbeat natural experience in a rural surrounding.
- Several scientific and empirical methods were adopted for an analysis of the geo-tourism potential of the identified sites. The various traverses of geological interest in the project area were recognised through early field work, survey data and information compiles from various sources, such as scientific articles, theses, thematic maps, and other related studies in the area.
- Several field work, to identify the traverses and rock exposures in the site area were conducted using toposheet (Open series map no. H44N10) to evaluate their geological attributes for their geo-educational potential. All locations with significant exposures were studied in detail and marked on map along with their GPS coordinates. These fieldworks were also utilized to identify and assess various heritage houses for converting them into home stays or low-cost accommodations with all amenities.
- Rock samples were collected and numbered with location code during the field works were analysed in laboratory for the petrological studies. Thin-section studies were conducted for petrography and micro-biota using petrological and stereo-zoom microscopes. Thematic maps like Geological, geoheritage, Geomorphological, Naula (ancient water reservoirs), Trail, Important recreational places in and around area were created using toposheet (1:50,000), Google earth imageries and freely available DEMs in ArcMap 10.3.
- Sketch, Photo and Video documentation was done for all important rock exposures to highlight the geological and structural features along with recreational points in the vicinity of site area.
- Many recreational locations such as river streams, waterfalls, ancient temples, and spring reservoirs (*Naulas*), archaeological monuments and trekking sites of general tourist interest were identified around the site to provide holistic travel experience.
- A guidebook with all sorts of valuable information has been prepared to make field work a memorable experience.

Second Phase: Development of affordable accommodation for the geo-tourist constituted the second phase of the project. After a field survey, several houses with heritage architecture (some being more than 100 years old) in the villages around the identified Geo-sites were found to be vacant or underutilised or under the care of a caretaker. Some of them requiring minor repair and renovation were selected to develop them as home stays/affordable accommodations. Necessary consent of the house owners was acquired after several consultations. Subsequently, a formal contract was signed on a legal document between the house owners and the project proponents mentioning obligations of both the parties. As per the project proposal 14 such units have been developed with modern facilities in these houses for the comforts of the Geo-tourists while maintaining their heritage value. Hospitality training programs have been conducted to train the village residents to provide a range of hospitality services to the tourists.

Third Phase: Any iota of success of a project in the rural areas is determined largely by the level of involvement of the local community. This therefore constitutes the third and crucial phase of the methodology. Considering the importance of the protection of Geoheritage and creation of additional employment in the target village cluster, extensive deliberations were held with the local Pradhan and villagers. Several awareness campaigns were conducted in the villages as well as in the nearby schools and inter-colleges to sensitize the local old and young population regarding the importance of the local geo-diversity and its protection (Fig. 6).

Some fossil collections were also displayed to the audience to highlight the fact that the Geo-sites contain evidence of life forms extending to several hundred million years ago. Brochures displaying biodiversity, various landforms, Geo-diversity, and disasters (both natural and man-made) were distributed to the local population. The development of home stay and its operations is also proposed to be based on a cooperative model where the villagers take over the responsibility of hospitality services and general operations of the home stays/affordable accommodation developed under the project. Self Help groups (SHGs) have also been formed in the

Fig. 6 Awareness program on Geoheritage/Geotourism in villages and schools in the vicinity of the project site. (**a–d**) awareness program on Geotourism among villagers in Kotyura, Kasoon, Patiya and Bhatgaon respectively. Rock mineral fossil show and Geoheritage Awareness campaign among schoolteachers and students in Kamleshwar Inter College (**e** and **g**), Inter College, Dinapani (**f**), Agriculture Inter College, Patiya (**h**)

village cluster to assist in augmentation of agricultural production which is a part of the holistic growth under the same project. Local administration has also been involved in the process of getting approvals for placing boards for protection of Geo-sites from indiscriminate mining or destruction as well as to highlight the geological features of the Geo-sites. An appeal has also been made to the district administration to declare the geo-site area as 'no mining zone'.

Site Progress

The present status of the development of Patiya cluster Geo-education cum Geotourism site is as under:

1. Identification of Geo-spots: About 19 locations all along the traverse have been marked in the map. At most of the geo-locations, outcrops information boards have been installed. Additionally, four information boards have been placed specifically for protection of the rocks after due approval from the district administration.
2. Excursion Guide: An excursion guide has been prepared incorporating exclusive information on all the 19 locations with photographs and geological details.
3. Student Visit: Seven student field trips have already been organised, immensely benefitting the undergraduate students (Fig. 7).

Fig. 7 Four images (**a–d**) depicting various field activities involving students and subject experts during different geoexcursion trips

Fig. 8 Photographs of some heritage and other houses developed as Homestays under Patiya village cluster project, displaying exterior, interior, and basic amenities available for the geotourists

4. Expert Visit: Several expert visits to the Geo-excursion site have been organized for their evaluation and valuable inputs (Fig. 7). These inputs will help in improving the descriptions of the exposures.
5. Development of Home stays/budget accommodations: As many as 14 units of home stays have been developed to provide accommodation to geotourists. Additional hospitality services like food, laundry, guides, transportation too have been arranged (Fig. 8).
6. Training Program: In order to provide hospitality services as per modern standards training programs have been provided by experts to several stakeholders.
7. International Conference: Two days International Conference on "Sustainable Geotourism to Preserve Geoheritage and Geodiversity in the Himalayas" in hybrid mode was successfully organised at Almora. The conference was well attended by national and international experts along with local administration and stakeholders (geology students and local community). Subsequently, as a part of the conference schedule, experts and participants were taken for a visit to this newly developed geo-excursion site to showcase the site where considerable time was spent by the experts who made valuable observations.
8. Publicity: As a part of policy an interactive website (https://himalayangeoexcursion.com) was developed for promotional activities and other social media accounts were also opened. In a much focused approach letters and brochure highlighting the site were also sent to the institutes offering courses in Geology.

6 Concluding Remarks

Development of Geo-excursion site using geological diversity of the place along with community participation as hospitality provider has resulted in a new form of geotourism serving primarily to geo-educational sector of the society in a significant way. Its true sustainable nature is well reflected in limited footfall of sensible

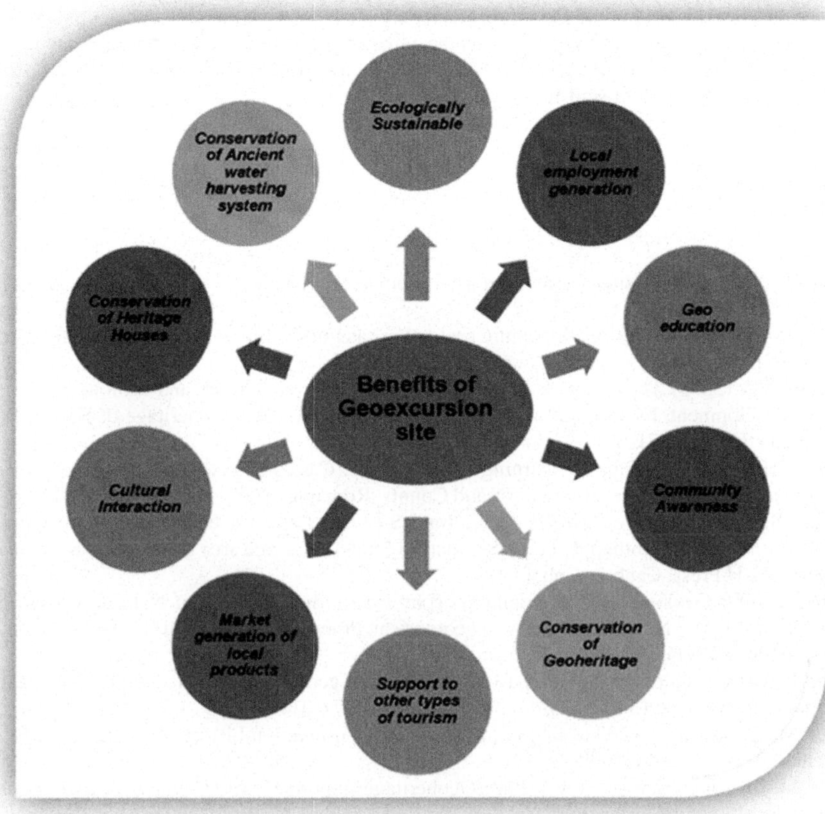

Fig. 9 Diagram illustrating the benefits of Geo-excursion site and its impact on different aspects of nature and material life of the local community

young students and researchers and serves as an ideal model of geotourism in fragile Himalayan environment strengthening people's participation in a beneficial manner (Fig. 9).

The developed site is first of its kind in India and has the potential to become a decisive catalyst in the efforts towards preservation of Geoheritage as well as its sustainability through local participation. A proactive involvement of the local community arising out of tourism activity managed by them would also instil a sense of pride besides ensuring its sustainability. In addition, it has the inherent component of promoting other forms of sustainable tourism like village-, eco-, cultural- and heritage- tourism. As a marker of initial success of the model it is worth mentioning here that several students' trips have already visited the site for field work which is gradually bringing a significant change in lives of the local community in a sustainable manner.

Acknowledgement The authors are thankful to Ministry of Environment, Forest and Climate Change and NMHS for funding the project under which the present research and action work was carried out. We are immensely grateful to the villagers of Patiya village cluster for providing access to their villages for conducting the study. We would also like to express our gratitude to the field attendants and support staff for their assistance.

References

Ahluwalia AD (2006) Indian geoheritage, geodiversity: geosites and geoparks. Curr Sci 91(10):1307–1307

Alexandrowicz Z (2006) Geoparks-nature protection category aiding the promotion of geotourism (Polish perspectives). Geoturystyka 2:3–12

Andrasanu A (2005) The Geopark-framework for research, education, and training in sustainable development. In: Proceedings of second conference on the Geoheritage of Serbia. ISBN 502.131-1.55, pp 171–175

Bâca I (2015) A curriculum for learning nature values. Case study: learning geodiversity from Bistriţa Ardeleană gorge (Bistriţa-Năsăud County, Romania). Geo J Tour Geosites 15(1):14–24

Bertzky B, Bertzky M, Worboys GL, Hamilton LS (2015) Earth's natural heritage. In: Worboys GL, Lockwood M, Kothari A, Feary S, Pulsford I (eds) Protected area governance and management. ANU Press, Canberra, pp 43–80

Brilha J (2009) Geological heritage and geoconservation in Portugal. In: Neto de Carvalho C, Rodrigues J (eds) New challenges with geotourism. Proceedings of the VIII European Geoparks Conference, Portugal, pp 31–35

Brilha J (2016) Inventory and quantitative assessment of geosites and geodiversity sites: a review. Geoheritage 8:119–134. https://doi.org/10.1007/s12371-014-0139-3

Brocx M, Semeniuk V (2007) Geoheritage and geoconservation-history, definition, scope and scale. J R Soc West Aust 90(2):53–87

Brocx M, Brown C, Semeniuk V (2019) Geoheritage importance of stratigraphic type sections, type localities and reference sites—review, discussion and protocols for geoconservation. Austral J Earth Sci 66(6):823–836. https://doi.org/10.1080/08120099.2019.1593242

Burek CV (2001) Non-geologists now dig geodiversity. Earth Heritage 16:21

Burek CV, Prosser CD (2008) The history of geoconservation: an introduction. Geol Soc Lond, Spec Publ 300(1):1–5

Célérier J, Harrison TM, Beyssac O, Herman F, Dunlap WJ, Alexander A, Webb G (2009) The Kumaon and Garhwal lesser Himalaya, India: part 2. Thermal and deformation histories. GSA Bull 121(9–10):1281–1297. https://doi.org/10.1130/B26343.1

Corte BD (2008) Geoparks, geosciences and environmental education. In: Proceedings 3rd International UNESCO Conference on Geoparks, Germany, pp 18–19

Cottrell SP, Vaske JJ, Shen F, Ritter P (2007) Resident perceptions of sustainable tourism in Chongdugou, China. Soc Nat Resour 20(6):512–513

Dowling RK (2013) Global geotourism–an emerging form of sustainable tourism. Czech J Tour 2(2):59–79

Dowling RK, Newsome D (2010) Geotourism a global activity. Good fellow Publishers Limited

Eder FW (2004) The global UNESCO network of Geoparks. In: Zhao X, Jiang J, Dong S, Li M, Zhao T (eds) Proceedings of the first international conference on Geoparks. Geological Publishing House, Beijing, pp 1–3

Eder W (2008) Geoparks- promotion of earth sciences through geoheritage conservation, education, and tourism. J Geol Soc India 72:49–154

Farsani NT, Coelho C, Costa C, Carvalho CN (2011) Geoparks & geotourism: new approaches to sustainability for the 21st century. Brown Walker Press, Florida

Fassoulas C (2003) Psiloritis Natural Park: capabilities for education and development of hinterland. In: NHM of Petrified Forest. Proceedings of 2nd International Symposium of Natural Monuments and Geological Heritage, Greece, pp 260–268

Frey ML (2003) Geopark Vulkaneifel: geopotential, touristic valorization, and sustainable development. In: Proceedings of the 2nd European Geoparks Network Meeting, Lesvos 3–7 October 2001, pp 29–44

Gray M (2003) Geodiversity. Valuing and conserving abiotic nature. Wiley, p 434

Gray M (2004) Geodiversity—valuing and conserving abiotic nature. Wiley, Chichester

Gray M (2013) Geodiversity: valuing and conserving abiotic nature, 2nd edn. Wiley Blackwell

Hose TA (2005) Geo-tourism—appreciating the deep time of landscapes. In: Novelli M (ed) Niche tourism: contemporary issues, trends and cases. Elsevier, Oxford, pp 27–37

Hose TA (2016) Three centuries (1670–1970) of appreciating physical landscapes. In: Hose TA (ed) Appreciating physical landscapes: three hundred years of geotourism, Geological Society Special Publication No. 417, London, pp 1–23

Hose TA, Vasiljević DA (2012) Defining the nature and purpose of modern geotourism with particular reference to the United Kingdom and south-East Europe. Geoheritage 4:25–43. https://doi.org/10.1007/s12371-011-0050-0

Ielenicz M (2009) Geotope, geosite, geomorphosite. The annals of Valahia university of Targoviste. Geogr Ser 9:7–22

INTACH (2016) A monograph on National Geoheritage Monuments of India. Indian National Trust for Art and Cultural Heritage(Natural Heritage Division), Delhi, p 146

Joyce EB (2010) Australia's geoheritage: history of study, a new inventory of geosites and applications to geotourism and geoparks. Geoheritage 2:9–56

Kale VS (2015) India's geo-wealth, geomorphological heritage and geomorphodiversity: assessment, classification and promotion. J Indian Geomorphol 3:10–26

Larwood J, Prosser C (1998) Geotourism, conservation and tourism. Geologica Balcanica 28:97–100

Mazumdar MK (2007) The 'geopark' initiative. Curr Sci 92:12

McKeever PJ, Narbonne GM (2021) Geological world heritage: a revised global framework for the application of criterion (viii) of the world heritage convention. IUCN, Gland. https://doi.org/10.2305/IUCN.CH.2021.12.en

Nekouie-Sadry B (2009) Fundamentals of geotourism: with emphasis on Iran. Samt Organization Publishing, Tehran, 220 p

Pande P, Kumar R, Sachdev R, Pant NC (2018) Geo-tourism in Kumaon region of Uttarakhand, India- a latent catalyst for business, employment and economic growth. Res Rev Int J Multidiscipl 3(3):383–394. ISSN: 2455-3085 (Online)

Prosser C (2002) Terms of endearment. Earth heritage 17: 12.13.Joint nature conservation committee, English nature, Scottish natural heritage and the countryside Council for Wales

Puniya MK, Patel RC, Pant PD (2019) Structural and thermochronological studies of the Almora klippe, Kumaun, NW India: implications for crustal thickening and exhumation of the NW Himalaya. Geol Soc Lond, Spec Publ 481:81–110. https://doi.org/10.1144/SP481-2017-74

Sharples C (2002) Concepts and principles of Geoconservation. Published electronically on the Tasmanian Parks & Wildlife Service website, pp 1–79. https://nre.tas.gov.au/Documents/geo-conservation.pdf

Shekhar S, Kumar P, Chauhan G, Thakkar MG (2019) Conservation and sustainable development of Geoheritage, Geopark, and Geotourism: a case study of Cenozoic successions of Western Kutch. India Geoheritage 11(4):1475–1488. https://doi.org/10.1007/s12371-019-00362-5

Stanley M (2000) Geodiversity. Earth Heritage 14:15–18

Stanley M (2004) Geodiversity–linking people, landscapes and their culture. In: Parkes MA (ed) Natural and cultural landscape–the geological foundation. In Royal Irish Academy, Dublin, pp 47–52

Swarna K, Biswas SK, Harinarayana T (2013) Development of Geotourism in Kutch region, Gujarat, India: an innovative approach. J Environ Prot 4:1360–1372

Geological Evolution and Geoheritage of Palaeoarchean Bundelkhand Craton, India

Vivek P. Malviya, H. Thomas, A. Krishnakanta Singh, and D. C. Meshram

Abstract The Bundelkhand craton is one of the oldest remnants of Earth's crust having recorded the geological evidence of the evolution of the primitive crust. The semicircular body extending over about 29,000 km^2 consists of (i) the highly deformed oldest rock (TTG and greenstone rocks) (3.59–3.20 and 2.7 billion years) (ii) the less deformed older rocks (K-rich granite and quartz reefs; ~2.5–2.0 billion years) and (iii) the undeformed younger rocks (different phases of mafic dykes) (~2.37, 2.0–1.7, 1.53, 1.46 and 1.1 billion years). Bundelkhand craton preserves some of the important geological features, which can be developed as geotourism sites, namely, Palaeoarchean granite-greenstone rocks (~3.59 Ga), Dhala impact structures, one of the oldest impact structures on Earth, orbicular granite, rare granitic rock, spectacular giant quartz reefs intruded in the entire craton and the Eparchaean unconformity (contact between Bundelkhand granite and Vindhyan Supergroup of rocks) Archean-Proterozoic boundary.

Keywords Bundelkhand craton · Vindhyan Supergroup · Orbicular granite · Archean-Proterozoic boundary

1 Introduction

Indian subcontinent has preserves variety of rocks, geological features, fossils, geological/tectonic boundaries and landscapes ranging from Paleoarchean to recent. These natural sites must be suitably delineated and protected preserves and developed as geoheritage sites. The development of geoheritage sites will be useful for the conservation of sites of geological importance, promote geosciences education

V. P. Malviya (✉) · H. Thomas · A. K. Singh · D. C. Meshram
Department of Applied Geology, Dr. Harisingh Gour Vishwavidyalaya,
Sagar, Madhya Pradesh, India
e-mail: vivek.geology@dhsgsu.edu.in

S. C. Tripathi et al. (eds.), *Geoconservation and Geotourism Potential of India*,
Society of Earth Scientists Series, https://doi.org/10.1007/978-3-031-81017-6_8

in the society and encourage geotourism in the area. In this connection several government organizations, geoscience societies and NGO, s are working collectively to develop several such sites. Presently development of the geoparks is also the major international theme implying geosciences application for inclusive growth of society and protection/conservation of unique geoheritage.

Indian peninsular shield was formed by the nucleation, accretion, and collision of such blocks, the northern (Bundelkhand and Aravalli) crustal block and the southern (Dharwar, Singhbhum and Bastar) crustal block separated by the Central Indian Tectonic Zone (Fig. 1a). These cratonic segments preserve the history of early evolution of continental crust in Indian shield (Radhakrishna and Naqvi 1986). Bundelkhand craton is one of the blocks that preserve records of the early evolution of Earth and vis-a-vis the evolution of the North Indian continental block. The growth rates and development of this continental block have changed over time. The oldest geological material ever found on Earth is 4.40 billion years old in detrital zircon, which is technically not a rock but a mineral grain recovered from sediment at Jack Hill, Slave Province, Western Australia (Wilde et al. 2001). The oldest known rock on Earth dated 4.03 billion years, is part of the Acasta Gneiss of the Slave craton in northwestern Canada (Bowring and Williams 1999). Technically the oldest known rock on Earth does not belong to Earth; instead, it is of extraterrestrial origin. Lunar rock samples collected during the Apollo mission dated 4.54 billion years, a few hundred million years older than our oldest terrestrial rocks. In January; 2020 astronomers reported that the oldest material found on Earth to date is silicon carbide particles from the Murchison meteorite, which has been determined to be 7 billion years old, about 2.5 billion years older than 4.54 billion years old Earth (Heck et al. 2020). However, scientists also explain that this age is based on theoretical models using assumptions with large uncertainties. The oldest reported rock in India is from Champua, Kendujhar district, Odisha from Singhbhum craton, which dates as old as 4.24 and 4.03 billion years (Chaudhuri et al. 2018). This article presents the general geological feature of the Bundelkhand cratonic block and the overall evolution of this part of Earth's continental crust also enumerate on inventory of geoheritage sites of Paleoarchean Bundelkhand craton recommends approaches for geoconservation.

The Bundelkhand craton not only preserves the evolutionary records of the formation of the North Indian continental block but also has several important geological features such as ~3.5 Ga old granite-greenstone rocks exposed in Mauranipur and Babina area, Dhala meteoritic impact site, the seventh oldest impact structure on Earth, the orbicular granite exposed in Pichhore area, Shivpuri district and Manpur area, Chitrakoot district, are one of the rare rocks on Earth, probably formed through nucleation around a grain in a cooling magma chamber due to rapid physical changes, the spectacular giant quartz reefs, are probably result of hydrothermal activity. The Eparchaeanun conformity, Archean-Proterozoic (~2.5 Ga) boundary exposed in the Chitrakoot district. These features are potential sites for geo-tourism and can be marked as geoheritage sites of India.

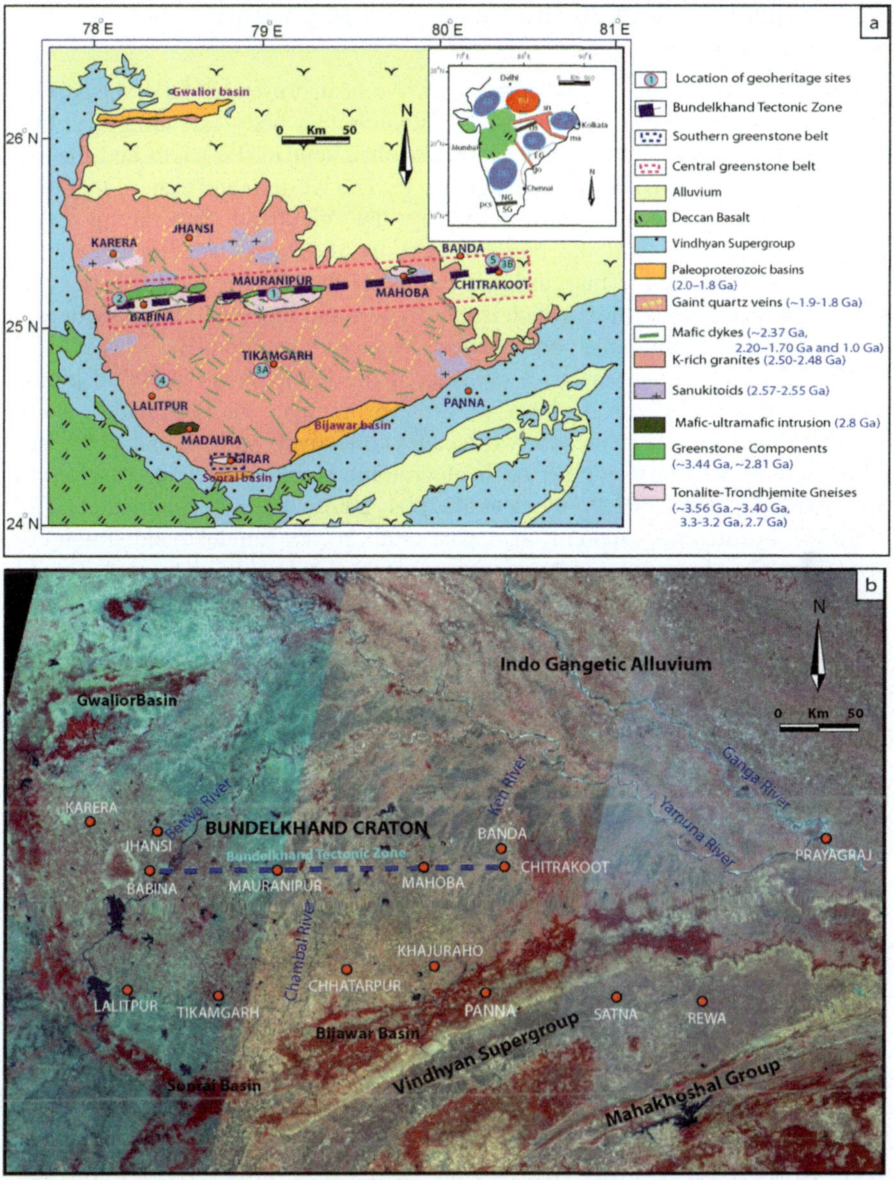

Fig. 1 (**a**) Geological map of Bundelkhand craton illustrating prominent lithologies, surrounding boundaries, and the location of proposed geoheritage sites (modified after Malviya et al. 2006). Inset tectonic map of peninsular Indian shield showing various tectonic elements and location of cratonic blocks (modified after Naqvi and Rogers 1987) Abbreviations: AR—Aravali Craton, BA—Baster Craton, BU—Bundelkhand Craton, DH-Dharwar Craton, SI-Sunghbhum Craton, sn: Son Narmada rift, ma: Mahanadi graben, go: Godavari graben, pcs—Palghat-Cauvery shear zone, cis—Central Indian suture, NG-Northern Granulite Terrain, SG-Southern granulite terrain, EG-Eastern Ghats terrain; (**b**) Landsat-7 (ETM+) satellite image of Bundelkhand craton and adjoining area showing marked linear features as giant quartz vein and mafic dykes within craton and also showing boundaries of the craton

2 General Geology of the Bundelkhand Craton

The Bundelkhand Craton is one of the oldest Archean crustal nuclei of the Northern Indian peninsular shield. It is exposed as semicircular body extending over about 29,000 km^2, bounded by the Son Narmada lineament in the south and the great boundary fault of Aravalli in the west (Fig. 1). The western and eastern margins of the Bundelkhand craton are covered by the Vindhyan Supergroup (Meso-Neoproterozoic age) and the northern part by the Indo-Gangetic alluviumand Himalayan rocks. Three marginal rift basins of the Paleo-proterozoic age were developed around the craton, which is the Bijawar and the Sonari basin to the south and the Gwalior basin to the north-western fringe of the craton (Basu 1986).

Bundelkhand craton was once considered the monotonous granitoid plutonic complex, dominantly of Archean to Proterozoic age. Various studies on the Bundelkhand craton from the last two decades collectively suggest that it has a similar evolutionary history as other Archean cratonic blocks worldwide (Sarkar et al. 1996; Mondal et al. 2002; Malviya et al. 2004, 2006, 2022a, b, Singh et al. 2019a, b, 2021a, b, references therein, Sensarma et al. 2021). Bundelkhand craton comprises mainly three litho-tectonic components, (i) Highly deformed Tonalite-Trondhjemite-Granodiorite (TTG)-greenstone components, (ii) Multi-phase granitic intrusion and associated quartz reefs (NNE-SSWto NE-SW trend) and (iii) Mafic dyke swarms (NW-SE trend dominantly). Based on the review of all the recent studies, the general stratigraphy of the Bundelkhand craton is given in Table 1. The detailed description of these three litho-tectonic components is as below.

2.1 Highly Deformed Tonalite-Trondhjemite-Granodiorite (TTG)-Greenstone Components

Tonalite–trondhjemite–granodiorite (TTG) series granitoid and greenstone belt assemblages of the craton represent thePalaeo-Mesoarchean crust and give typical Achaean craton status to Bundelkhand. Mainly three phases of TTG gneisses (Table 1 and Fig. 2a) are emplaced in the craton. The first phase (TTG-I) are the oldest rocks of Bundelkhand Craton, these rocks show pre- to syngenetic evidence with greenstone belt rocks (Fig. 2b–d; Singh et al. 2019b). They were emplaced around 3.59–3.55 Ga and are reported from the Babina and Mauranipur areas (Kaur et al. 2014; Saha et al. 2016). However, the second phase (TTG-II) is the most abundant TTG gneisses are emplaced during 3.3–3.2 Ga, which also occurs around the Babina, Mauranipur, Mahoba area (Fig. 1a; Mondal et al. 2002; Kaur et al. 2016; Batuk Joshi et al. 2017; Singh et al. 2021a, b). The third and last phases (TTG-III) are emplaced during ~2.7 Ga Verma et al. (2016) and are reported from the Babina area. The protolith of TTG gneisses is interpreted to have been formed in subduction-related tectonic settings at the garnet stability P-T field (TTG-I), plagioclase stability field (TTG-II) and intermediate depth (TTG-III). Malviya et al. (2022a) suggest

Table 1 General stratigraphy of Bundelkhand Craton; Modified after Malviya et al. (2022a, b)

Palaeoproterozoic marginal Basin (Bijawar—Sonrai—Gwalior)	2.0–1.8 Ga	Absar et al. (2009), Chakraborty et al. (2015)
Giant Quartz vein	1.9–1.8 Ga	Pati et al. (2007), Slabunov and Singh (2022).
----------------------------------Intrusive contact--------------------------------		
Mafic-ultramafic dykes swarm	2.37 Ga, 2.2–1.7 Ga, 1.53 and 1.46 Ga 1.1 Ga	Pradhan et al. (2012), Radhakrishna et al. (2020) Raju et al. (2024)
Sanukitoids and K-rich granites	2.59–2.48 Ga	Mondal et al. (2002), Kaur et al. (2016), Batuk Joshi et al. (2017), Singh et al. (2021a, b)
----------------------------------Intrusive contact--------------------------------		
Madaura mafic-ultramafic complex	3.20–2.57 Ga (inferred age from field relation), 2.8 Ga	Ramiz et al. (2018), Mohanty et al. (2019), Slabunov et al. (2024)
Felsic Volcanics	2.80 and 2.54 Ga	Singh and Slabunov (2015), Slabunov and Singh (2019)
----------------------------------Intrusive contact--------------------------------		
Tonalite-trondhjemite gneises (TTG-III)	~2.7 Ga	Verma et al. (2016), Singh et al. (2021a, b)
Tonalite-trondhjemite gneises (TTG-II)	3.40–3.20 Ga	Mondal et al. (2002), Kaur et al. (2016), Singh et al. (2019a), Singh et al. (2021a, b)
----------------------------------Intrusive contact--------------------------------		
Central Bundelkhand Greenstone complex (Babina-Mauranipur-Mahaba Green stone belt) Pillowed ultramafic-mafic volcanics(Basaltic komatiite-Boninitic); Felsic volcanics, BIF, Quartzite, volcaniclastic metasediments)	3.8–3.5 Ga (Nd model ages); (~3.44 Ga, Sm-Nd isochron age) (3.2 Ga,2.8 Ga, 2.5 Ga and 2.1 Ga, **Metamorphic ages**)	Malviya et al. (2004, 2005, 2006, 2013); Singh et al. (2019b), Saha et al. (2011) Raza et al. (2021), Malviya et al. (2022b)
Southern Bundelkhand Green stone complex (Girar-Rangaon-Berwar greenstone belt) **Rajaula Formation** (Quartz feldspar biotite gneiss, chlorite schist and meta-basalts) **Berwar Formation**(Basal quartz conglomerate, fuchsite-bearing quartzite, grey to green chloritic shale and banded magnetite-hematite quartzite	3.4–3.2 Ga (Zr-age from Quartzite) 3.29 Ga (Nd-Model Ages) 2.98 Ga (Nd-Model Ages)	Singh and Slabunov (2015), Singh and Slabunov (2016), Slabunov et al. (2017), Hiloidari et al. (2021)
----------------------------------Intrusive contact--------------------------------		
Tonalite-trondhjemite gneises (TTG-I)	3.59–3.40 Ga	Kaur et al. (2014, 2016), Saha et al. (2016), Nasipuri et al. (2019), Malviya et al. (2022a)

Fig. 2 (**a**) Field photograph showing mafic enclave within TTG gneisses in Saprar river section, Mauranipur; (**b**) Field photograph of isoclinal folded BIF showing F1 deformational feature in Babina greenstone belt, (**c**) Field photograph showing deformed quartzite from the Mahoba greenstone belt; (**d**) Field photographs showing pillowed structure in mafic volcanic from the Mauranipur greenstone belt

that the source material for the TTG-I was rutile-free amphibolite and/or Eoarchean tonalite, whereas for the TTG-II and the TTG-III it was an enriched mafic source. The oldest TTGs (3.59–3.55) of the Bundelkhand Craton were formed by the partial melting of pure oceanic crust in the garnet stability field, while the moderate-high degree of partial melting of the enriched mantle source at the plagioclase stability field to generate the younger TTGs (3.3–3.2 Ga), with minor assimilation of older felsic crustal components during its ascent (Kaur et al. 2016; Malviya et al. 2022a).

Two greenstone belt complexes can be recognized in the Bundelkhand craton (Fig. 1; Table 1), namely the (i) Central Bundelkhand greenstone complex (Babina-Mauranipur-Mahoba greenstone belts) along Bundelkhand Tectonic Zone (BTZ; Malviya et al. 2006) and (ii) Southern Bundelkhand greenstone complex (Rungaon-Girar greenstone belt; Singh and Slabunov 2015). Central Bundelkhand greenstone complex (Babina-Mauranipur-Mahoba greenstone belts) are generally E-W trending granite-greenstone terrane in the central part of the craton around ~250 km long stretches from the Chitrakoot, Mahoba to Babina via Mauranipur (Fig. 1; Malviya et al. 2006, 2022a, b) and follow the trend of Bundelkhand Tectonic Zone (BTZ; Pati et al. 2007). The central Bundelkhand complex consists of pillowed basalt, mafic-ultramafic rocks of basaltic komatiitic-bainitic affinity, and less often metasedimentary and Banded Iron Formation (BIF; Fig. 2b–d; Malviya et al. 2006; Singh and Slabunov 2015; Singh et al. 2019b; Malviya et al. 2022b). Bundelkhand BIF of central greenstone belt is classified as Algoma type. They occur along the BTZ and provide strong evidence of the existence of proto-oceanic basins in the central part of the Bundelkhand craton. Geochemical data suggests different sediment sources, multiple proto-basins, and varied physicochemical conditions for BIF deposition during the Archean time (Malviya et al. 2013, 2022b). These greenstone belts are remnants of a supra-subduction crust of the island arc that has been tectonically accreted to the continent during the arc-continent collision and alignment of the greenstone belt along the BTZ is Archean suture (Malviya et al. 2006).

The Southern Bundelkhand greenstone belt (Rungaon—Girar greenstone belt) is located in the southern part of the craton and consists generally of meta-sedimentary rocks namely quartzite and BIF. Quartzite is represented by fuchsite and hematite-bearing quartz arenite and lesser quartz pebble conglomerates that have been subjected to low-grade metamorphism. BIF is represented by thinly-bedded quartz–hematite (±magnetite) rocks. The BIF of the Girar greenstone Belt differs from the rest in relatively high U, Ag, and Bi, low Al_2O_3, CaO, Ti, and $\sum$REE, and extremely high variations in SiO_2 and Fe. These BIFs seem to have formed in a basin with high hydrothermal activity, far from terrigenous source areas. BIF is fairly rich in Cr and Ni, poor in Zr, Hf, Ba, Th, Sr, Yb, and Lu, and displays a distinct positive Eu anomaly (Eu/Eu* = 1.14–2.46; Singh and Slabunov 2016). Zircons from quartzite give two U-Pb ages: 3432 ± 9.7 Ma and 3252 ± 6.4 Ma. The Sm-Nd isotope study of quartzite from the Girar belt shows that the TDM is 3.29 Ga. This TDM correlates well with the U-Pb ages of zircon and indicates that the continental crust in the Girar area began to form in the Paleoarchaean (3.4–3.2 Ga; Singh and Slabunov 2016). Recently Hiloidari et al. (2021) proposed that metavolcanics of Kurrat—Girar—Badwar greenstone belt formed in a modern style Mesoarchean extension-subduction geodynamic regime.

Madawara ultramafic complex lies at the southern fringe of the craton and includes altered and metamorphosed dunite, peridotites, pyroxenite, gabbro, and diorite (Thomas et al. 2019). Diorite occurs as an E-W trending body (Table 1). Mohanty et al. (2019) suggest an arc environment that is similar to Alaskan-type intrusion. Ramiz et al. (2018) suggested that these rocks formed in a subduction zone setting. Based on field cross-cutting relation <3.2 to >2.57 age is assigned for these rocks Ramiz et al. (2018).

2.2 Multiphase Granitic Intrusion and Quartz Reef

Close to the Archaean–Proterozoic boundary, a vast amount of geochemically variable high—K calc-alkaline granitoid plutons (Figs. 1a and 3a, b) intruded into the TTG—greenstone protocontinent during a narrow period between 2.59 and 2.48 Ga. The granitoid can be divided into a low-silica high-magnesium group indicating

Fig. 3 (**a** and **b**) Field photograph of various K-rich granite variants from Jhansi and Lalitpur area; (**c** and **d**) Photographs showing regional field occurrence of giant quartz vein from Lalitpur and Mauranipur area; (**e**) Field photograph show intrusive nature of younger mafic dykes from LadaPahar area, Mahoba

enriched mantle origin and mantle-crust interactions (sanukitoid granitoids and Closepet-granites), and a high-silica low-magnesium group pointing to pure crustal melting (monzogranite with low-HREE and low-Eu subgroups; Batuk Joshi et al. 2017). The variable geochemical signatures of the high-K calc-alkaline granitoid can be best explained by (a) mantle wedge enrichment both by subduction and subsequent slab breakoff-related fluid/melt activity and (b) fluid-assisted partial melting in the mantle and crust at different depths at a convergent continental margin. Batuk Joshi et al. (2017) reported whole-rock geochemistry and Nd isotope data for some specific granitoids (sanukitoids and anatectic granites) to explain their temporal evolution and genetic relationship. These sanukitoids originated in a subduction-related tectonic setting and were emplaced between 2.58 and 2.50 Ga, which is contemporaneous with high-K anatectic granite magmatism (partial melting of pre-existing TTGs). Isotopic study of sanukitoids reveals that whole-rock εNdt values are mostly negative (−3.6 to −1.6) suggesting a significant role of the crustal component in the generation of these granitoids. Possibly the mixing of metasomatized melts from the mantle with anatectic granitic melts, followed by homogenization at shallow crustal level, resulted in the formation of sanukitoids (Singh et al. 2019a). Colleps et al. (2021), recently proposed crystallization age of granitoids between 2.4 and 2.3Ga based on apatite U-Pb ages suggests broad-scale exhumation of the Bundelkhand Craton through the mid-crustal depths following amalgamation and felsic magmatism and stabilization of craton by ~2.2 Ga.

The giant quartz veins in the Bundelkhand craton are observed as spectacular landforms on the regional scale (Figs. 1a and 3c, d). These trends, in general, in NNE-SSW to NE-SW direction and rarely exhibit E-W orientation and can be very easily seen on Landsat satellite imagery (Fig. 1b). Giant quartz veins (GQVs; earlier referred to as 'quartz reefs') represent gigantic Precambrian (~2.0 Ga; Pati et al. 2007) silica-rich fluid activity in the Central Indian shield. Their outcrop widths vary from ≤1 to 70 m and pervasively extend over tens of kilometers along the strike over the entire craton. They show imprints of strong brittle to ductile-brittle deformation, and in places are associated with base metal and gold incidences, and pyrophyllite-diaspore mineralization. The petrography, geochemistry, and fluid inclusion studies of giant quartz veins reveal their emplacement due to tectonically controlled polyphone hydrothermal fluid activity (Pati et al. 2007). The fluid inclusion data show the presence of bi-phase (H_2O–CO_2), hypersaline, and moderate temperature (100–300 °C) primary stretched fluid inclusions suggesting a possible hydrothermal origin for the Mo-bearing quartz occurring within variably deformed different granitoid variants of Archean Bundelkhand craton (Pati et al. 2014). The formation of a giant quartz vein swarm is associated with the deformation of the Bundelkhand Craton lithosphere during 1.9–1.8 Ga (Slabunov and Singh 2022).

2.3 *Mafic Dyke Swarms*

The Bundelkhand mafic dyke swarm consists of NW-SE trending mafic dykes that generally intrude the eastern and southwestern margins of the craton (Figs. 1a and 3e). One such nearly E-W trending dyke, more than 10 km long, forms a prominent ridge near Mahoba, called Mahoba dyke. The mafic dykes are low- to medium-K subalkaline, olivine- to hypersthene-normative tholeiites, predominantly reflecting continental-type magmatism emplaced in an extensional tectonic regime. The tightly constrained bulk SiO_2 content of the mafic dykes suggests that despite their varied spatial and temporal evolution, they are essentially co-magmatic. The intrusion of mafic magma into the ABC marks a periodic crustal dilation possibly due to a plume-related magmatism during the Proterozoic time (Pati et al. 2008a).

Paleomagnetic and geochronological data from the mafic dykes intruding in Bundelkhand craton (BC) is significant in constraining the position of India at 2.0 and 1.1 Ga (Pradhan et al. 2012). The three distinctive paleomagnetic directions obtained in this study for the mafic dyke swarms suggest three different pulses of emplacement within the craton. The precise U–Pb isotopic zircon age of ca1980 Ma for the NW–SE trending dyke and ca1113 Ma for the ESE–WSW trending great dyke of Mahoba support paleomagnetic analysis (Pradhan et al. 2012). Recently, based on geochemistry Radhakrishna et al. (2020) divided these dyke swarms into two groups: Group-I is intruded ~2.37 Ga, 2.18–2.20 Ga, while group-II intruded during 1.98 Ga.

3 Metamorphic Evolution

The metamorphic evolution of Bundelkhand Craton is not well understood; very limited information about metamorphism is present. The central Bundelkhand greenstone belt displays a poly-metamorphic evolution pattern. Saha et al. (2011) reported Mesoarchean (ca. 2.8 Ga) ultra-high-pressure metamorphism event from the Babina area, and the amphibole-plagioclase-quartz assemblage from TTG from the same area suggest P-T estimates in the range of ~6.5–8.5 kbar and 630–720 °C, Nasipuri et al. (2019). Metamorphic studies of central Bundelkhand greenstone terrane Sibelev et al. (2021) suggest at least three stages of metamorphic processes, the first stage is ca 2.8 Ga eclogite facies, the second is ca.2.7 Ga amphibolite facies with retrograde branch and the third and last is, ca. 1.9–1.8 Ga prehnite-pumpellyite facies. The metamorphic study of garnetiferous BIF from the Mauranipurgreenstone belt reveals the P-T estimate with temperature varying between 450 and 500 °C and pressure ranging between 0.1 and 0.2 GPa indicating a lower amphibolite condition and also given three metamorphic age ~3.2 Ga, 2.5 Ga and 2.1 Ga based on monazite ages (Raza et al. 2021). A systematic and comprehensive metamorphic study of various litho-units along with geochronological data is very much required to properly understand the metamorphic evolution of Bundelkhand craton.

4 Summary of Geochronological Data of Bundelkhand Craton

The geochronological data of the Bundelkhand craton was very little till the last decade. Many geochronological studies published in the last decade provide very useful information about the evolution of craton through space and time. A summary of available geochronological data location-wise is presented in Fig. 4 for better understanding. Based on age data, the highly deformed Tonalite-Trondhjemite-Granodiorite (TTG) can be divided into three stages ~3.5 Ga, 3.3–3.2 Ga, and ~2.7 Ga. The greenstone mafic-ultramafic volcanic components' age implies that their formation is pre-syngenetic with TTG magmatism. Sm-Nd Modal ages suggest Eo-palaeoarchean age Malviya et al. (2005), while Sm-Ndisochronage implies ~3.44 Ga age for it (Singh et al. 2019b). Statistical analysis of 33 age data of sanukitoids and K-rich granite implies that this episode started at 2.59 Ga and ended at

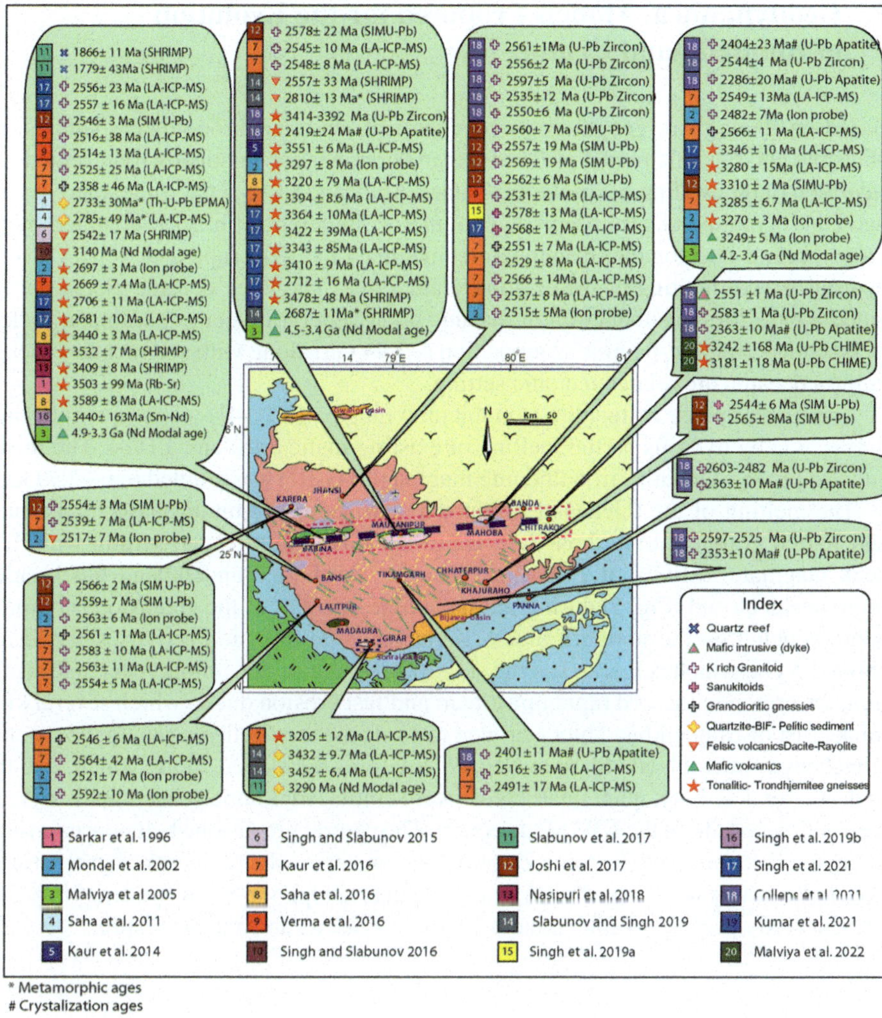

Fig. 4 Summary of location-wise geochronological data of various litho-units of Bundelkhand Craton

2.48 Ga with the peak at 2.55 Ga. The hydrothermal activity-induced quartz reef intrusion occurs during ~2.0 Ga. The mafic dyke activity is dated as 1.7–2.0 Ga and 1.1 Ga, while U-Pb ages from separated zircon from mafic dyke samples give ~2.5 Ga ages, Colleps et al. (2021). While Radhakrishna et al. (2020) divided these dyke swarms into two groups Group-I is intruded ~2.37 Ga, 2.18–2.20 Ga, while group-II intruded during 1.98 Ga. The age data of the quartz reef and mafic dyke is creating confusion and needs more high-quality age databases to separate these two events.

5 Geodynamical Models Proposed for the Evolution of Bundelkhand Craton

Based on geological, geophysical, geochemical, and geochronological studies, a plausible geodynamic model of Bundelkhand craton is schematically represented as a cartoon diagram (Fig. 5). In the Eo-Palaeo Archean (4.0–3.5 Ga) time, there were two continental blocks (northern and southern Bundelkhand block) separated by a proto ocean. The tonalite-trondhjemite magma (TTG-I) erupted at ~3.5 Ga by hydrous partial melting of the existing mafic crust. The thick oceanic plateau provided a base for the subduction of oceanic crust generating mafic-ultramafic rocks of Basaltic komatiite-bainitic composition (~3.44 Ga) along with sedimentary rocks in intra oceanic arc related tectonic setting.

The second phase of tonalite-trondhjemite magma (TTG-II) was erupted during 3.3–3.2 Ga by hydrous partial melting of existing mafic crust and TTG-I. The third and last phase of tonalite-trondhjemite magma (TTG-III) was erupted (ca. ~2.7 Ga).

The melting of the thickened subducted slab generated multi-phase sanukitoid and K-rich granite during 2.59–2.48 Ga in the subduction-related setting with the peak magmatic activity at 2.55 Ga. When the northern Bundelkhand block side proto ocean closed (Fig. 5) during the same time, the part of the greenstone component abducted on the northern block. K-rich granitic intrusion activity stopped at about 2.5 Ga, which is marked as the stabilization age of the craton. After stabilization, the craton witnessed rapid upliftment and fast erosion due to which several km thick crust of Bundelkhand block eroded towards the then Tethyan sea. In the extensional environment NNE-SSW to NE-SW fractures developed, along which, due to hydrothermal activity, giant quartz reefs were emplaced. Plume-related mafic dykes were emplaced along the NW-SE direction. The central greenstone belt now exposed along the Bundelkhand Tectonic Zone, which preserves the remnants of greenstone belt rocks, was formed in early Archean time in a subduction-related setting (Fig. 5). Present study proposes Bundelkhand Tectonic Zone as an Archean suture.

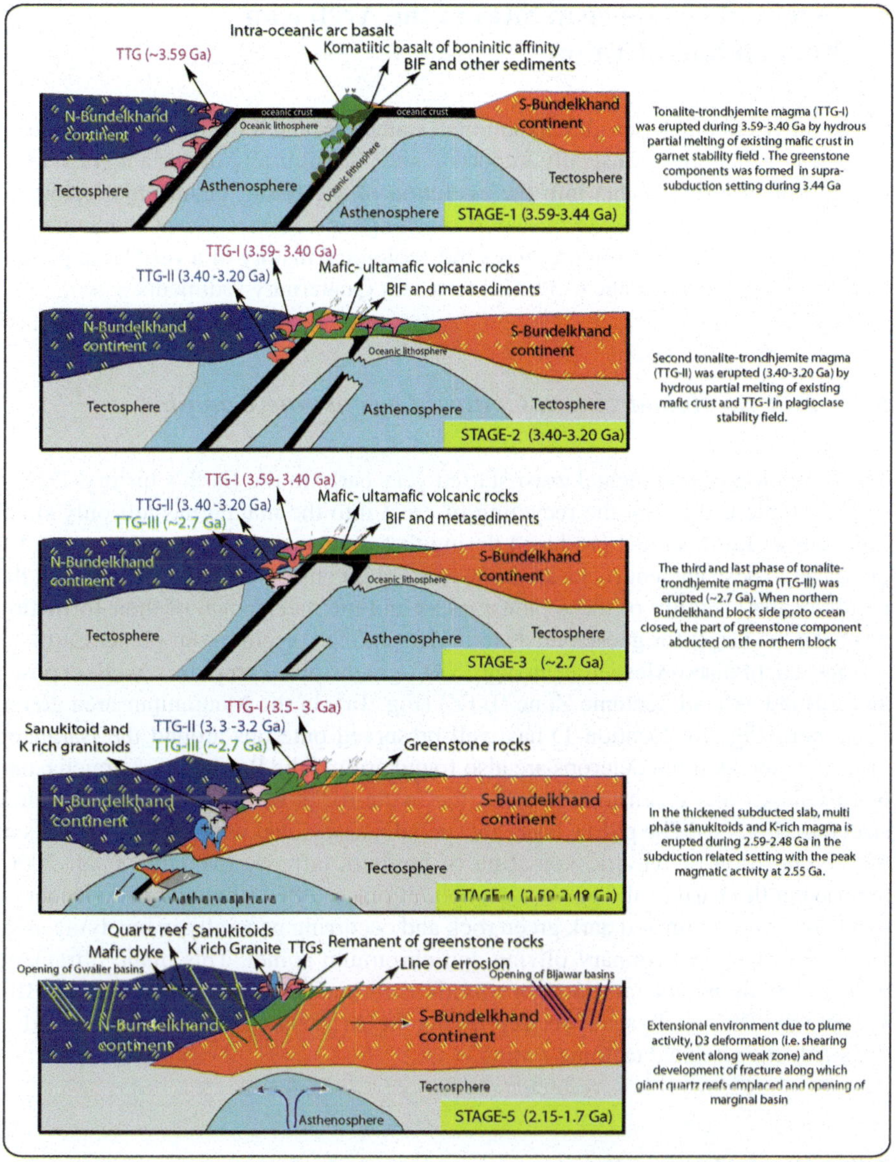

Fig. 5 Cartoon diagram depicting plausible geodynamic model for geodynamical evolution Bundelkhand craton based on the available geochemical and geochronological database

6 Potential Geotourism Sites in the Archaean Bundelkhand Craton

Geoheritage encompasses global, national, statewide, and local features of geology, at all scales that are intrinsically important sites or culturally important sites offering information or insights into the evolution of the Earth; or into the history of science, or that can be used for research, teaching, or reference Brocx and Semeniuk (2007). Bundelkhand craton preserves the geological history of a very large period, from the Palaeoarchean age (~3.6 Ga) to recent Quaternary sediments.

6.1 Palaeo-Mesoarchean Central Greenstone Complex

The formation of continental crust started very early in the Earth's history. Due to plate tectonic processes, the recycling of crust into the mantle has left only small remnants of Earth's crust produced during the early years of Earth's formation for preserves. These ancient rocks are known as the greenstone belt. Understanding the nature of the protolith of these oldest rocks and the mechanism of their formation has been the target for geoscientists to understand the origin of our planet Earth.

One suchPalaeo-Mesoarchean age TTG-greenstone complexes well exposed along Bundelkhand Tectonic Zone (BTZ) (Fig. 1a). In the Mauranipur area greenstone belt (Fig. 1a; location-1) has well-preserved outcrops around the Baragaon village, whereas some outcrops are also found around the Raspahari, Kuraicha, and Roni villages this greenstone belt has the mafic rocks with well-preserved pillow structure (Fig. 2d). The pillow lavas are spatially associated with ultramafic rocks of the composition of basaltic komatiite of boninitic affinity (Malviya et al. 2006) exposed at the base of the greenstone belt, at places, the ultramafic rocks occur as extensively serpentinized dark green rock and occurring generally in low lying area, with well-preserved primary olivine and chromium spinel. This mafic ultramafic unit is overlain by volcanoclastic metasediments and Banded Iron Formation (BIF; Fig. 2b–d). The whole sequence represents the Archean ophiolitic belt, formed in the subduction-related tectonic setting.

6.2 Dhala Impact Structure

The Dhala impact structure covers nearly 64 km^2 in parts of Shivpuri district, Madhya Pradesh in north-central India (Fig. 1a; location-2 and Fig. 6a, b). The Dhala structure was previously considered a "Cauldron Structure" (Bhattacharya et al. 2011; Gaur et al. 2016) of some type of crypto-volcanic origin. Several studies on the area, especially by Pati et al. (2008b), Roy et al. (2017), Li et al. (2018), Keerthy et al. (2019), Pati et al. (2019), Agarwal et al. (2020) and Singh et al.

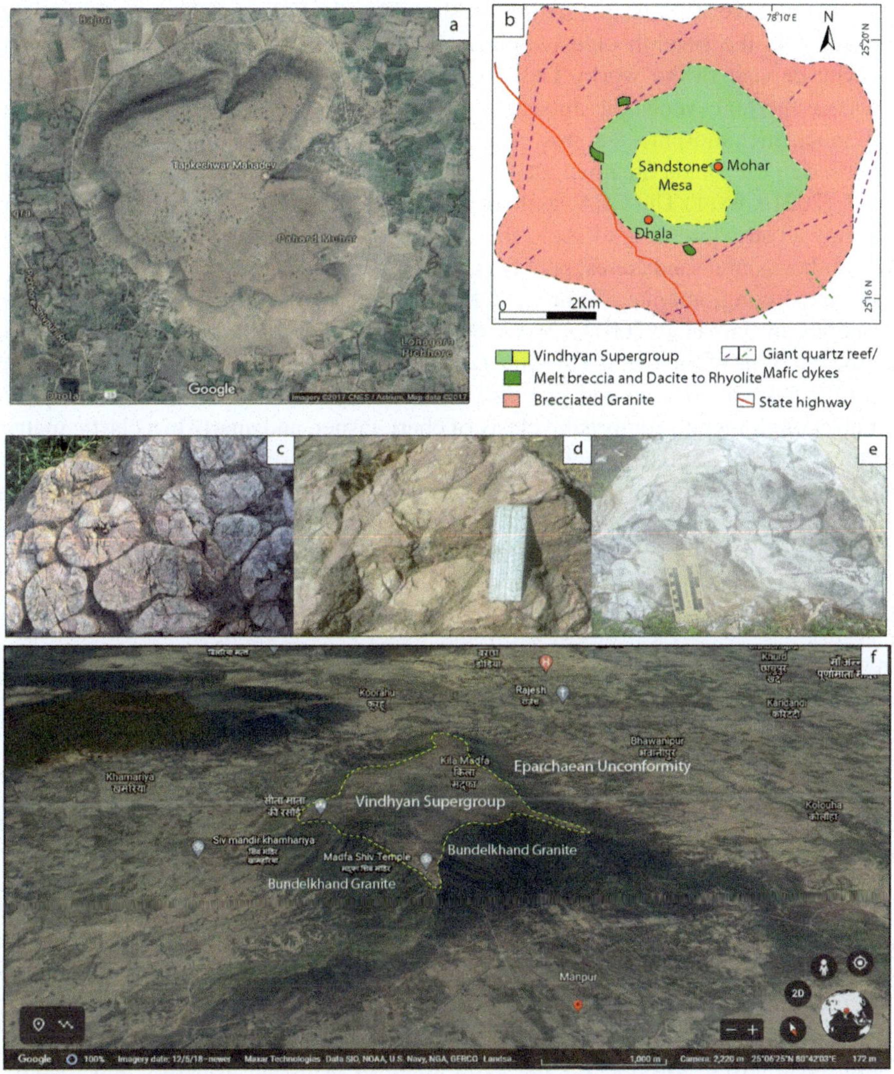

Fig. 6 (**a**) Google Earth satellite image showing the boundary of Dhala impact structure; (**b**) Geological map of Dhala impact structure after Li et al. (2018); (**c**) Field photo showing unicellular orbicules from orbicular granite exposed near Kutawali village in Pichhore tehsil in Shivpuri district, M.P.; (**d**) Field photo show unicellular orbicules with the nucleus from orbicular granite exposed near Kutawali village in Pichhore tehsil in Shivpuri district, M.P.; (**e**) Field photo show orbicular granite exposed near Manpur village, Chitrakoot district, UP.; (**f**) Google Earth satellite image showing Archean–Proterozoic boundary (Eparchaean unconformity) near Manpur village, Chitrakoot district, UP

(2021a, b) gave it the status of one of the oldest known impact structures in the world. It is the seventh oldest impact structure with possibly the oldest known suevite deposit in the world. Despite the deep level of erosion and post-impact tectonic-thermal events, the impactites are exceedingly well preserved, with nearly all shock metamorphic features. The three distinct morpho-lithological domains exposed in the Dhala area include the following (Fig. 6b) Jain et al. (2001), Singh et al. (2021a, b) (1) The CEA, with an area of about 5 km^2, is located off-center for the impact structure. It is offset to the south and comprises an alternating sequence of sub-horizontal sandstones, pebbly sandstones, and shale, with small exposures of lateritized conglomerate (Semri and Kaimur sedimentary rocks of the Vindhyan Supergroup; Jain et al. 2001). (2) A topographically nearly flat area of about 16 km^2 in the environs of the CEA. This area comprises mainly sub-horizontal, intercalated purplish-brown siltstone and greenish-white sandy siltstone, with lenses of polymictic conglomerate comprising clasts of chert, jasper, and quartz in a clastic matrix (Jain et al. 2001; Singh et al. 2021a, b) (3) The third morpho-lithological domain is defined by a ring of more than 200 outcrops of essentially monomict breccia. The breccia is predominantly composed of granitoid clasts besides a minor component of quartz vein-derived clastic material.

6.3 Orbicular Granite

The orbicular structures are preserved in the Neoarchean Bundelkhand Granites, which also happen to be the major lithology of the region. In Bundelkhand Craton, orbicular granite is reported from two locations. The first site is located SSW of the Kutawali village (25°07.274′N,78°09.897′E; Fig. 1a; location 3A) in Pichhore Tehsil, Shivpuri district of Madhya Pradesh. Both single and multi-shelled orbicules have circular rims of biotite at their margins and are 20–25 cm in diameter (Fig. 6c, d).

In the Pichhore area, the aesthetically beautiful concentric flower-like structures are preserved in the medium-grained pink porphyritic granite outcrop lying in the middle of a farming field and therefore thought to have been dangered. This outcrop is an ideal site to develop for the geotourism of orbicular structure and rapakivi texture in post-kinematic granitoids. It occurs as a near-circular outcrop of approximately 30 m × 40 m in size within the coarse-grained granite phase. The exposure is a composite of two types of orbicules, one is large potash feldspar rich and another is smaller in size and richer in white feldspar than pink feldspars. The orbicular are either closely packed with a tangential contact or are separated by enough groundmass material so that their forms are fairly regular. The density and the size of orbicules vary significantly within the outcrop. However, the average orbicules groundmass ratio is about 70:30. The rock is granular, pink-colored biotite granite and consists of orthoclase, perthite, quartz, oligoclase, and some biotite as the major constituent minerals.

Another location of orbicular granite occurs about 600 m SSW of Manpur (25°8′18″N, 80°42′54″E; Fig. 1a; location-3B) Chitrakoot district of Uttar Pradesh. This body is occurring within deformed tonalite-trondhjemlte-granodiorite (TTG) gneiss. The exposed outcrop (Fig. 6e) size is about 54 m^2. However, its true dimensions could not be deciphered due to the alluvial cover on the northeastern and northern sides of the outcrop, Pati (2004).

This site of orbicular granite of the Bundelkhand craton is unique in the sense that it provides an extremely good site to researchers for comparative studies on orbicular and rapakivi granites, as well as for the studies on the theory of magma mingling and study of post-syntectonic granitoid of Precambrian time.

6.4 *Spectacular Giant Quartz Reefs*

The giant quartz reefs in the Bundelkhand craton are a spectacular feature. They occur as a ridge with a gentle slope on either side cropping out over the surrounding plain and are easily identified in topographic maps and LANDSAT satellite imageries (Figs. 1b and 3c–d). The giant quartz reefs (generally NNE-SSW to NE-SW trend) emplacement takes place during 1.9–1.8 Ga in the entire craton (Pati et al. 2007; Slabunov and Singh 2022). The outcrop pattern of giant quartz veins is variable, but most veins are linear with a very high aspect ratio and local sigmoidal geometry. The maximum outcrop length of the giant quartz veins is more than 60 km (Basu 1986). However, they can be traced along the strike for over 100 km as discontinuous pinch and swell type outcrops with the average length of individual outcrops exceeding 1 km. Their width ranges from <1 m to 70 m with very few exceeding 20 m on the profile section. Debris on either side of the giant quartz veins tends to increase the apparent width up to as large as 300 m noticed at Bar, Lalitpur district. In general, the GQVs are thicker in the middle and taper on both ends along their length (Pati et al. 2007, 2014; Slabunov and Singh 2022).

6.5 *Archean-Proterozoic Boundary: Eparchaean Unconformity*

The Archean-Proterozoic boundary also known as "Eparchaean Unconformity" is placed at ~2500 Ma and marks possibly the most dramatic change in Earth's history. The 'Eparchaean Unconformity' is a major discontinuity of stratigraphic significance. The Great Oxidation Event (GOE) took place between 2.45 and 2.32 Ga as part of an Archean-Proterozoic transition. Apart from geological proxies of atmospheric oxygenation, such as banded iron formation (BIF) abundance and paleosol mineralogy, isotope chemostratigraphy provides the most powerful tool for studying the GOE and establishing the boundary. The time gap of ~500 Ma during the

Archean-Proterozoic boundary preserves lots of information related to changes in atmospheric conditions Gaucher and Frei (2018). In geological history, the Archean-Proterozoic boundary is marked as a major unconformity and preserved very rarely and provides us a unique opportunity to study the changes that occurred during this transition period.

This geoheritage site is located around Manpur village (25°06′25″N, 80°42′03″E) of Chitrakoot district, Uttar Pradesh (Fig. 1a; location 5). Along with this location, several other flat top hills expose this unconformity (Fig. 6f). In this location the lower Vindhyan Supergroup sedimentary rocks are overlain by ~2.5 Ga Bundelkhand granite. This unconformity forms a specific boundary between the 'Proterozoic Vindhyan supergroup rocks and the Archaean Bundelkhand granite. The Vindhyan Supergroup was formed between 1700 and 650 million years ago, which essentially comprises sedimentary rocks, whereas the Achaean rocks which comprise various granites and gneisses are more than 2500 million years old. This time gap of more than 700 Ma is quite significant in the creation of the Earth's crust. The unconformity is very significant to the geologists studying the area as this gap may provide an answer to many unresolved geological problems.

7 Geotourism Potential and Conservation of Geoheritage Site of the Area

Bundelkhand region is very rich in cultural heritage and has many famous places such as Chitakoot famed in Hindu mythology and the epic Ramayana, Jhansi the gateway of Bundelkhand region linked with the legend Rani LaxmiBai and her majestic fort, the Orchha temples of sixteenth century AD build by Bundela in Niwari District and famous temples of Khajuraho built by Chaldela King in the ninth century are the main tourist attraction. The development of the proposed geotourism site will be of value addition to the tourists visiting the region.

References

Absar N, Raza M, Roy M, Naqvi SM, Roy AK (2009) Composition and weathering conditions of Paleoproterozoic upper crust of Bundelkhand craton, Central India: records from geochemistry of clastic sediments of 1.9 Ga Gwalior Group. Precambrian Res 168(3–4):313–329

Agarwal A, Kumar S, Joshi G, Agarwal KK (2020) Evidence for shock provides insight into the formation of the central elevated area in the Dhala impact structure, India. Meteoritics Planetary Sci 55(12):2772–2779

Basu AK (1986) Geology of parts of the Bundelkhand Granite Massif, Central India. Geol Survey India Spec Publication 117:61–124

Batuk Joshi K, Bhattacharjee J, Rai G, Halla J, Ahmad T, Kurhila M, Heilimo E, Choudhary AK (2017) The diversification of granitoids andplate tectonic implications at the Archaean–Proterozoic boundary in the Bundelkhand Craton, Central India. Geol Soc Lond 449(1):123–157

Bhattacharya D, Joshi GB, Sharma R (2011) Uranium mineralisation associated with felsic volcanism at Mohar, Shivpuri district, Madhya Pradesh. J Geol Soc India 78(1):57–62

Bowring SA, Williams IS (1999) Priscoan (4.00–4.03Ga) orthogneisses from northwestern Canada. Contrib Mineral Petrol 134:3–16

Brocx M, Semeniuk V (2007) Geoheritage and geoconservation - history, definition, scope, and scale. J R Soc Western Austr 90(2):53–87

Chakraborty PP, Pant NC, Paul PP (2015) Controls on sedimentation in Indian Palaeoproterozoic basins: clues from the Gwalior and Bijawar basins, central India. Geol Soc Lond Memoirs 43(1):67–83

Chaudhuri T, Wan Y, Mazumder R, Ma M, Liu D (2018) Evidence of enriched, Hadean mantle reservoir from 4.2–4.0 Ga zircon xenocrysts from Paleoarchean TTGs of the Singhbhum Craton, Eastern India. Sci Rep 8(1):7069

Colleps CL, McKenzie NR, Sharma M, Liu H, Gibson TM, Chen W, Stockli DF (2021) Zircon and apatite U-Pb age constraints from the Bundelkhand craton and Proterozoic strata of central India: Insights into craton stabilization and subsequent basin evolution. Precambrian Res 362:106286

Gaucher C, Frei R (2018) The Archean-Proterozoic boundary and the great oxidation event. Chemostratigrap Across Major Chronol Bound 30:33–45

Gaur VP, Survanshi H, Shrivastava SK, Nambiar KV (2016) Stratigraphy and correlation of Mesoproterozoic rocks associated with Mohar cauldron, Shivpuri district, Madhya Pradesh. J Geol Soc India 8(5):603–608

Heck PR, Greer J, Kööp L, Trappitsch R, Gyngard F, Busemann H, Maden C, Ávila JN, Davis AM, Wieler R (2020) Lifetimes of interstellar dust from cosmic ray exposure ages of presolar silicon carbide. Proc Natl Acad Sci 117(4):1884–1889

Hiloidari S, Satyanarayanan M, Singh SP, Bhutani R, Subramanyam KSV, Sarma DS (2021) Evidence for Mesoarchean subduction in Southern Bundelkhand Craton, India: geochemical fingerprints from metavolcanics of Kurrat-Girar-Badwar Greenstone Belt. Geochemistry 81(3):125787

Jain SC, Gaur VP, Srivastava (2001) Recent find of a Cauldron structure in Bundelkhand Craton. Geol. Surv. India. Spec Publ 64:289–297

Kaur P, Zeh A, Chaudhri N (2014) Characterisation and U Pb- Hf isotope record of the 3.55 Ga felsic crust from the Bundelkhand Craton, northern India. Precambrian Res 255:236–244

Kaur P, Zeh A, Chaudhri N, Eliyas N (2016) Unravelling the record of Archaean crustal evolution of the Bundelkhand Craton, northern India using U–Pb zircon–monazite ages, Lu–Hf isotope systematics, and whole-rock geochemistry of granitoids. Precambrian Res 281:384–413

Keerthy S, Vishnu CL, Li SS, Reshma N, Praveen MN, Oommen T, Singh SP, Sajinkumar KS (2019) Reconstructing the dimension of Dhala Impact Crater, Central India, through integrated geographic information system and geological records. Planetary Space Sci 177:104691

Li SS, Keerthy S, Santosh M, Singh SP, Deering CD, Satyanarayanan M, Praveen MN, Aneeshkumar V, Indu GK, Anilkumar Y, Sajinkumar KS (2018) Anatomy of impactites and shocked zircon grains from Dhala reveals Paleoproterozoic meteorite impact in the Archean basement rocks of Central India. Gondwana Res 54:81–101

Malviya VP, Arima M, Pati JK, Kaneko Y (2004) First report of metamorphosed basaltic pillow lava from central part of Bundelkhand Craton, India: an Island arc setting of possible Late Archean age. Gondwana Res 7S:1338–1340

Malviya VP, Arima M, Pati JK (2005) Island arc related Archean mafic volcanism from Bundelkhand craton, Central India – an implication from Nd Model ages. Joint meeting of Earth and Planetary Sciences, Makuhari, Japan

Malviya VP, Arima M, Pati JK, Kaneko Y (2006) Petrology and geochemistry of metamorphosed basaltic pillow lava and basaltic komatiite in the Mauranipur area: subduction related volcanism in the Archean Bundelkhand craton, Central India. J Mineral Petrol Sci 101(4):199–217

Malviya VP, Arima M, Pati JK (2013) Petrochemical and geochemical study of Banded Iron Formation (BIF) from Archaean Greenstone Belt of Bundelkhand craton. 3rdPrecambrian

Continental Growth and Tectonism, Jhansi, India, Abstract volume International Association for Gondwana Research Conference Series 16, 99–100
Malviya VP, Arima M, Verma SK, Pati JK, Suzuki K, Dora ML, Singh VK, Shareef M (2022a) Geochemistry and U–Pb CHIME Ages of Tonalite–Trondhjemitic–Granodioritic (TTG) Gneiss from the Central Bundelkhand Craton, India: implication for the presence of Paleoarchean crust from Easternmost exposed boundary of the Craton. In: Geochem treasures petrogenetic processes. Springer Nature Singapore, Singapore, pp 207–241
Malviya VP, Arima M, Verma SK, Pati J, Shareef M, Dora M, Singh P, Shukla M (2022b) Petrology and geochemistry of the Palaeo-Mesoarchean Banded Iron formations (BIFs) from the central Bundelkhand greenstone belts, Bundelkhand Craton, India: source characteristics and depositional environment. Geol J 57(8):3292–3312
Mohanty N, Singh SP, Satyanarayanan M, Jayananda M, Korakoppa MM, Hiloidari S (2019) Chromian spinel compositions from Madawara ultramafics, Bundelkhand Craton: implications on petrogenesis and tectonic evolution of the southern part of Bundelkhand Craton,Central India. Geol J 54(4):2099–2123
Mondal MEA, Goswami JN, Deomurari MP, Sharma KK (2002) Ion microprobe 207Pb/206Pb ages of zircons from the Bundelkhand massif, northern India: implications for crustal evolution of the Bundelkhand–Aravalli protocontinent. Precambrian Res 117(1–2):85–100
Nasipuri P, Saha L, Hangqiang X, Pati JK, Satyanaryanan M, Sarkar S, Bhandari A, Gaur Y (2019) Paleoarchean crustal evolution of the Bundelkhand Craton, north Central India. Earth's Oldest Rocks 31:793–817
Pati JK (2004) Single-shelled orbicular granitoids of possible Archaean age from Bundelkhand Granitoid Complex (BUGC). Geol Soc India 64:223–226
Pati JK, Patel SC, Pruseth KL, Malviya VP, Arima M, Raju S, Pati P, Prakash K (2007) Geology and geochemistry of giant quartz veins from the Bundelkhand Craton, central India and their implications. J Earth Syst Sci 116:497–510
Pati JK, Raju S, Malviya VP, Bhushan R, Prakash K, Patel SC (2008a) Mafic dykes of Bundelkhand craton, Central India: field, petrological and geochemical characteristics. Indian dykes: Geochem Geophys Geochron, pp 547–569
Pati JK, Reimold WU, Koeberl C, Pati P (2008b) The Dhala structure, Bundelkhand craton, Central India-eroded remnant of a large Paleoproterozoic impact structure. Meteoritics Planetary Sci 43(8):1383–1398
Pati JK, Panigrahi MK, Chakarvorty M (2014) Granite-hosted molybdenite mineralization from Archean Bundelkhand craton molybdenite characterization, host rock mineralogy, petrology, and fluid inclusion characteristics of Mo-bearing quartz. Journal of Earth System Science. 123:943–958
Pati JK, Poelchau MH, Reimold WU, Nakamura N, Kuriyama Y, Singh AK (2019) Documentation of shock features in impactites from the Dhala impact structure, India. Meteoritics Planetary Sci 54(10):2312–2333
Pradhan VR, Meert JG, Pandit MK, Kamenov G, Mondal MEA (2012) Paleomagnetic and geochronological studies of the mafic dyke swarms of Bundelkhand craton, central India: implications for the tectonic evolution and paleogeographic reconstructions. Precambrian Res 198:51–76
Radhakrishna BP, Naqvi SM (1986) Precambrian continental crust of India and its evolution. J Geol 94:145–166
Radhakrishna T, Tomson JK, Chandra R, Ramakrishna C (2020) Geochemistry of NW-SE trending Palaeoproterozoic mafic dyke intrusions in the Bundelkhand Craton, India and subcontinental lithospheric mantle processes. Precambrian Res 351:105956
Raju S, Bodas MS, Anshu R, Neogi S (2024) Proterozoic mafic dyke swarms of Bundelkhand Craton, North India: a connection to Columbia supercontinent. Geol J 59(7):2098–2117
Ramiz MM, Mondal MEA, Farooq SH (2018) Geochemistry of ultramafic-mafic rocks of the Madawara Ultramafic Complex in the southern part of the Bundelkhand Craton, Central Indian Shield: implications for mantle sources and geodynamic setting. J Geol 53:1–23

Raza MB, Hifzurrahman, Nasipuri P, Saha L, Pati JK, Alfimova NA, Champati AK (2021) Phase relations and in-situ U-Th-Pb total monazite geochronology of Banded Iron Formation, Bundelkhand Craton, North-Central India, and their geodynamic implications. Int J Earth Sci 111:287–315

Roy M, Pandey P, Kumar S, Parihar PS (2017) Mineral shock signatures in rocks from Dhala (Mohar) impact structure, Shivpuri district, Madhya Pradesh, India. J Earth Syst Sci 126:1–13

Saha L, Pant NC, Pati JK, Upadhyay D, Berndt J, Bhattacharya A, Satynarayanan M (2011) Neoarchean high-pressure margarite–phengitic muscovite–chlorite corona mantled corundum in quartz-free high-Mg, Al phlogopite–chlorite schists from the Bundelkhand craton, north central India. Contrib Mineral Petrol 161:511–530

Saha L, Frei D, Gerdes A, Pati JK, Sarkar S, Patole V, Bhandari A, Nasipuri P (2016) Crustal geodynamics from the Archaean Bundelkhand Craton, India: constraints from zircon U–Pb–Hf isotope studies. Geol Mag 153(1):179–192

Sarkar A, Paul DK, Potts PJ (1996) Geochronology and geochemistry of Mid-Archean trondhjemitic gneisses from the Bundelkhand craton, central India. Recent Res Geol 16:76–92

Sensarma S, Matin A, Paul D, Madhesiya AK, Sarkar G (2021) Evolution of a crustal-scale silicic to intermediate tectono-magmatic system: the ~2600–2300 Ma Bundelkhand granitoid, India. Precambrian Res 352:105951

Sibelev OS, Slabunov AI, Singh VK, Mishra S (2021) Metamorphism of the Central Bundelkhand Greenstone Complex of the Bundelkhand Craton, Indian Shield and its geodynamic setting. In: Shandilya AK, Singh VK, Bhatt SC, Dubey CS (eds) Geological and geo-environmental processes on Earth. Springer Natural Hazards, Springer, Singapore, pp 143–154

Singh VK, Slabunov A (2015) The Central Bundelkhand Archaean greenstone complex, Bundelkhand craton, central India: geology, composition, and geochronology of supracrustal rocks. Int Geol Rev 57:1349–1364

Singh VK, Slabunov A (2016) Two types of Archean supracrustal belts in the Bundelkhand craton, India: geology, geochemistry, age and implication for craton crustal evolution. J Geol Soc India 88:539–548

Singh PK, Verma SK, Singh VK (2019a) Geochemistry and petrogenesis of sanukitoids and high-K anatectic granites from the Bundelkhand Craton, India: Implications for late-Archean crustal evolution. J Asian Earth Sci 174:263–282

Singh PK, Verma SK, Moreno JA, Singh VK, Malviya VP, Oliveira EP, Mishra S, Arima M (2019b) Geochemistry and Sm-Nd isotope systematics of mafic-ultramafic rocks from the Babina and Mauranipur greenstone belts, Bundelkhand Craton, India: implications for tectonic setting and Paleoarchean mantle evolution. Lithos 330:90–107

Singh AK, Pati JK, Sinha R, Reimold WU, Prakash K, Nadeem M, Dwivedi S, Mishra D, Dwivedi AK (2021a) Characteristic landforms and geomorphic features associated with impact structures: Observations at the Dhala structure, north-central India. Earth Surface Processes Landforms 46(8):1482–1503

Singh PK, Verma SK, Singh VK, Moreno JA, Oliveira EP, Li XH, Malviya VP, Prakash D (2021b) Geochronology and petrogenesis of the TTG gneisses and granitoids from the Central Bundelkhand granite-greenstone terrane, Bundelkhand Craton, India: implications for Archean crustal evolution and cratonization. Precambrian Res 359:106210

Slabunov AI, Singh VK (2019) Meso–Neoarchaean crustal evolution of the Bundelkhand Craton, Indian Shield: new data from greenstone belts. International Geology Review, 61(11):1409–1428. https://doi.org/10.1080/00206814.2018.1512906

Slabunov AI, Singh VK (2022) Giant quartz veins of the Bundelkhand craton, Indian shield: new geological data and U-Th-Pb age. Minerals 12(2):68

Slabunov A, Singh VK, Joshi KB, Li X (2017) Paleoarchean zircons from quartzite of South Bundelkhand Supracrustal Complex: origin and implications for crustal evolution in Bundelkhand Craton, Central India. Curr Sci 112:794–801

Slabunov AI, Egorova SV, Singh VK (2024) Mesoarchaean mafic–ultramafic Ikauna layered intrusion, Bundelkhand craton, India: geology, U–Th–Pb age (SHRIMP) and correlation. Int Geol Rev 1–13

Thomas H, Verma CB, Jain A, Ackerman L (2019) Spinel chemistry from madawara mafic ultramafic complex, Lalitpur-district, Uttar Pradesh, India. J Nepal Geological Soc 58:53–59

Verma SK, Verma SP, Oliveira EP, Singh VK, Moreno JA (2016) LA-SF-ICP-MS zircon U–Pb geochronology of granitic rocks from the central Bundelkhand greenstone complex, Bundelkhand craton, India. J Asian Earth Sci 118:125–137

Wilde SA, Valley JW, Peck WH, Graham CM (2001) Evidence from detrital zircons for the existence of continental crust and oceans on the Earth 4.4 Gyr ago. Nature 409(6817):175–178

Deccan Volcanic Province: When Lava Flooded the Indian Peninsula

Vivek S. Kale

Abstract The Deccan plateau is a prominent geomorphic feature of central and western India exposing the thick stacks of subhorizontal basaltic lava flows, occasionally interlayered with thin interflow horizons of paleosols or freshwater sediments. This voluminous outpouring of basaltic lava during the passage of Indian plate over the Réunion hotspot had a dramatic global environmental impact contributing to the terminal-Cretaceous mass extinction. The numerous fossil localities and dinosaurian nesting sites from the fringes of this province; its volcanological architecture and signature landscape makes the Deccan Volcanic Province a major geoheritage region hosting a wide spectrum of geotourism locations. The Deccan basalt is recognised by IUGS International Commission on Geoheritage as one of the first global list of 'heritage stones' in 2022.

Keywords Deccan trap · Basaltic lava · Terminal-cretaceous mass extinction · Cretaceous—Paleogene boundary

1 Introduction

The Deccan Volcanic Province (DVP) is one of the largest continental flood basalt provinces (CFBPs) of the world (Bryan and Ernst 2008). This >2000 m thick pile of subhorizontal basaltic flows occupies a contiguous exposed area of around 500,000 km^2 (Fig. 1) Large parts of this basaltic pile were eroded away or down-faulted below the Arabian Sea, leaving behind less than half its original size. The step-like profile of the hillslopes on this pile of basaltic flows (Inset: Fig. 1) gives its name derived from Sanskrit word "*Dakshin*" meaning south and Swedish word "*Trappa*" meaning stairway. Other associated igneous rocks (including rhyolites, tuffs, alkali basalts, phonolites, carbonatites and intrusive mafic dykes) constitute a

V. S. Kale (✉)
Advanced Center for Water Resources Development and Management, Pune, India

S. C. Tripathi et al. (eds.), *Geoconservation and Geotourism Potential of India*, Society of Earth Scientists Series, https://doi.org/10.1007/978-3-031-81017-6_9

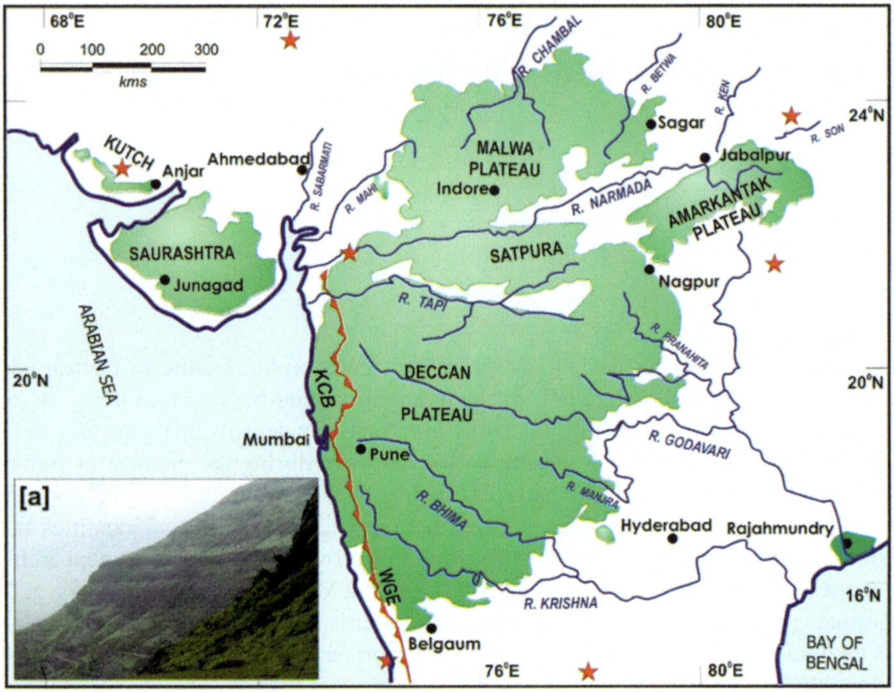

Fig. 1 Outline of the present exposures (green) of the Deccan Traps in Peninsular India and the geographic subprovinces of the Deccan Volcanic Province (DVP). The red stars represent exposures of mafic igneous activity that appear to have preceded the flood basalts (after Kale et al. 2020a). Inset (**a**) is a photograph of the step-like topographic profile typically encountered across the DVP. KCB = Konkan Coastal Belt; WGE = Western Ghats Escarpment

small fraction of this huge volume of basaltic magmatism (Kale 2020; Basu et al. 2020; Krishnamurthy 2020). The preserved basalts in the DVP represent a volume of around 3×10^5 km^3 by conservative estimates (Kale et al. 2020a), while Jay and Widdowson (2008) had earlier computed a possible volume of 13×10^5 km^3.

The DVP terrain is characterized by large flat plateaus (Deccan, Malwa, Amarkantak) fringed by steep escarpments (Fig. 2) and incised narrow river valleys with cascading streams (Dole et al. 2022) creating innumerable scenic locations that host popular hill stations and tourist spots, due to their scenic beauty. The eruptive history of the Deccan Traps straddling the Cretaceous—Paleogene (K-Pg) Boundary. That witnessed a major faunal turnover in Earth history (including the dinosaurian extinction and emergence of mammalian life) makes it a critical region for geological investigations. The wide spectrum of volcanological features hosted by the DVP and the environmental impact of this voluminous volcanism add to the flavor of it being a major geoheritage region.

2 Lava Dispersal

The phenomenal uniformity in physical appearance on account of similarities in the (chemical and mineralogical) composition of the tholeiitic basalt lava flows from most of the CFBPs makes them extremely difficult to classify. More than 90% in the DVP display a deviation of <1% from the weighted averages in major and minor elements (Kale et al. 2020b). Flow-by-flow mapping in several CFBPs has shown that individual flows occupy areas exceeding 1000 km^2. West (1959) had traced lava flows around Sagar across more than 100 km laterally. The Yakima basalt flow covering 18,000 km^2 is an example. A recent study (Shandilya et al. 2020) traced the continuity of a lava flow-field across an area >37,000 km^2 in the western DVP. The continuity of the lava flow fields in the western DVP across tens of kilometers is evident (Fig. 2b) along the >500 km long Western Ghats Escarpment (WGE).

The individual flows/flow-fields in the DVP have thicknesses varying from ~2 m to more than 50 m (Fig. 2c); with a majority having thickness of 8–17 m (Self et al. 2021). Based on their lateral spread, this indicates that the volume of lava erupted during single eruptions in the Deccan may have varied from ~2.8 km^3 to >30,000 km^3.

Fig. 2 (**a**) Subvertical scarps exposing the subhorizontal stack of basaltic lava flows in the WGE. Note the deeply incised valleys and flat-topped hills with a step-like profile. (Loc: Varandha Ghat, Pune district, Maharashtra). (**b**) Continuously traceable lava flows (with thickness of a few tens of metres) across many tens of kilometers exposed in the steep west-facing escarpment of the WGE that towers over the Konkan Coastal plains on the right side in the photo (Loc: Amba Ghat section of Ratnagiri district, Maharashtra). (**c**) A > 50 m thick lava flow exposed in the Ankai Fort hill (near Manmad in Dhule district, Maharashtra) occurring as an isolated mesa surrounded by vertical cliff faces in the heartland of the Deccan Plateau

Fig. 2 (continued)

Such eruptions spanned a few years or decades, analogous to the volumetric rates of extrusion recorded during recent eruptions in Hawaii, Iceland, Italy, etc. and the computed rates in other CFBPs. It is estimated that the eruptions in the Deccan Province occurred at an average rate of 6–10 km^3/year (Self et al. 2021); resulting in piling up of the lava into a stack at the rates a few 1000 m/10^6 years (Kale and Pande 2022).

The spread of such large and long continental lava flows is attributed to a low viscosity of the lava emplaced rapidly on the surface with very high effusion rates and is aided by topographic slopes (Keszthelyi and Self 1998; Self et al. 1998; Keszthelyi et al. 1999). Tested and further refined in recent volcanic eruptions and spread of lava in Hawaii, Etna, Iceland, etc., this model has limitations when applied to the subaerial lava flows from continental flood basalts. The cooling and degassing of the spreading lava would lead to a progressive incremental viscosity making the lava sluggish. The subhorizontal disposition of the substrate on which most continental flood basalts rest poses additional problems for explaining large aerial spreads and their driving momentum. An alternative model that enables emplacement (often in multiple pulses) of the lava 'endogenously' under the insulating cover of an early-formed crust offers an efficient mode of transfer of lava (even while retaining its fluidity) in a flow-field that may remain active across several decades. Such inflated sheets can then occupy large areas without requiring a slope to enable their lateral spreading since the successive pulses of lava provide the requisite forward momentum (Walker 2009; Self et al. 2014; Kale et al. 2021).

The cooling and degassing history of the lava during and after it has spread leads to a wide spectrum of lava morphologies and stacking architectures of the lava flows in the CFBPs (including the DVP). This is a keenly studied topic in volcanism across the globe (e.g. Duraiswami et al. 2008; Dietterich et al. 2017; Tarquini et al. 2019; Mittal et al. 2021; Hyman et al. 2022) not only for scientific enquiry, but more so because of its implications on volcanic hazards and extra-terrestrial crustal studies.

The insulated slow cooling of the viscous erupted lava (after it stops moving) enables development of polygonal columnar joints (called entablature or colonnades) in the core of the lava flows and the basic dykes intruding them. Such columnar jointed basalts can achieve spectacular dimensions (akin to the Giant's causeway in Ireland); making them apt candidates as geoheritage sites, like the Potla dyke (Fig. 3a), Bharudpura stone wall (Fig. 3b), the Singhana war-bonnet (Fig. 3c) from Madhya Pradesh and the Chinchani multitiered columnar joints (Fig. 3d). Such columnar jointed basalts can easily be excavated, making them readily available sources of construction material in this province. They have been used for construction of many heritage buildings and forts across the DVP. This has led to their recognition by the apex global body (IUGS International Commission on Geoheritage) in the first listing of geoheritage stones released in 2022. Such locations deserve to be tagged as heritage sites and be statutorily protected from excavations or damage beyond their present status urgently.

Fig. 3 (**a**) Spectacular columnar jointed Potla dyke (date at 66.6 ± 0.5 Ma by Patil Pillai et al. 2022) in Dewas district Madhya Pradesh (see: Joshi and Gadgil 2024). (**b**) Giant subvertical columnar joints in the core of thick lava flow (from Kalisindh (Manpur) Formation of Malwa Group (dated at 66.3 ± 0.1 Ma by Tholt et al. 2023), exposed in the banks of stream near Bharudpura—Chauki (22°22′06″N, 75°29′0″E), Dhar district, Madhya Pradesh. (**c**) Singhania War Bonnet (22°12′0″N, 74°57′13″E), in the core of a lava flow from the Maheshwar (Narmada) Formation of Malwa Group, Dhar district, Madhya Pradesh; developed due to radial arrangement of columnar jointing. (A flow in the vicinity has been dated at 66.4 ± 0.1 Ma by Tholt et al. 2023). (**d**) Chinchani (randomly oriented) Multitiered Columnar Joints in the core of lava flow from Indrayani Formation (66.2 ± 0.05 Ma: Schoene et al. 2019) of Sahyadri Group exposed downstream of the Ghod dam (18°39′55″N, 74°29′17″E) on the boundary of Pune and Ahmednagar districts, Maharashtra

Fig. 3 (continued)

Fig. 3 (continued)

3 Hotspots, Flood Basalts and Sedimentary Basins

CFBPs are generally linked with the passage of the continental crust over 'Hotspots' are zones of the Earth's lithosphere, which are heated by narrow streams of superheated mantle (called plumes) rising upwards from the Earth's core-mantle boundary. The mantle plumes cause melting of the mantle around the plume-heads (called the sub-crustal lithospheric mantle (SCLM)) creating large volumes of magma that then rises by buoyancy. This magma commonly feeds the spreading ocean-floor basalts enabling the growth of the oceanic crust. In cases such hotspots occur below continental crust, the SCLM melts become the primary magmatic source of the mafic lavas in the CFBPs. Investigations into the primary magmatic source of the Deccan lavas are an evolving subject, but there is no ambiguity about the relationship between the DVP and Réunion hotspot.

Geochronology of the trail of volcanics and the paleo-position of the Indian Plate demonstrate unequivocally that the inception of the RSBP and spreading of the Eastern Indian Ocean are linked to the activity of the Kerguelen hotspot which has remained active in the last 130 million years (Coffin et al. 2002; Kale 2020; Zhang et al. 2023). On the west, the DVP is a result of the passage of the Indian plate over the Réunion hotspot in the Indian Ocean (Fig. 4). The journey of the Indian plate that was previously a part of the Gondwana supercontinent, till its collision with the Eurasian supercontinent is punctuated with rifting, continental break-up, hotspot volcanism, ocean expansion, CFB eruptions, and collision tectonics (Valdiya 2016). The impact of the passage of the Indian plate across the Réunion hotspot including

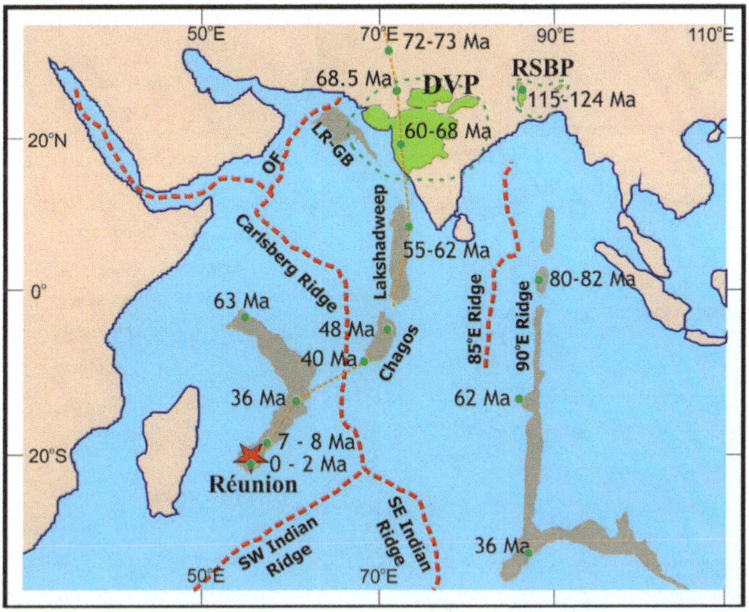

Fig. 4 Tectonic elements in the Indian Ocean (after Kale 2020: Fig. 8.19) depicting Deccan Volcanic Province (DVP) and Rajmahal—Sylhet—Bengal Province (RSBP) in the Indian subcontinent, mid-oceanic ridges (thick, red dashed line) and volcanic island chains and oceanic plateaus (with light brown shading). OF = Owen Fracture; LR-GB = Laxmi Ridge—Gop Basin. The trails of volcanic activity (marked by green dots with ages in Ma) connecting the RSPB with Kerguelen hotspot (located further south of the map area) and DVP with Reunion hotspot (red star) is shown

its separation history from Madagascar and the evolution of the western Indian ocean (Arabian Sea), upward doming and tilting (Sangode et al. 2022) are recorded in the stack of basaltic flows in the DVP and represent an important subject of study from the tectonic modelling of the history of the Indian plate since the late Cretaceous.

Fundamental differences in the mechanism of emplacement of basaltic lavas differentiate the continental flood basalts from the ocean-floor basalts. Although they share a common primary magmatic source. The presence of the continental crust over the magma chambers of the CFBPs (a) impedes its upward escape to the surface, enabling a higher degree of fractionation, (b) holds the upwelling magma in subcrustal chambers before expelling on the surface, (c) contaminates the magma with components that were not a part of its original constitution; and (d) cools more slowly during surface subaerial extrusion since the atmospheric temperature (>25 °C) is much higher than that on the ocean-floor (<5 °C).

Several sedimentary basins evolved on the Indian plate (Fig. 5) during the pretrappean Cretaceous period (145–66 Ma), such as the Kutch—Cambay basins (on the west) and the Bengal, Razole, Cauvery basins (on the east), besides those fringing the contemporary Tethyan Ocean in the north. During its northward flight (from

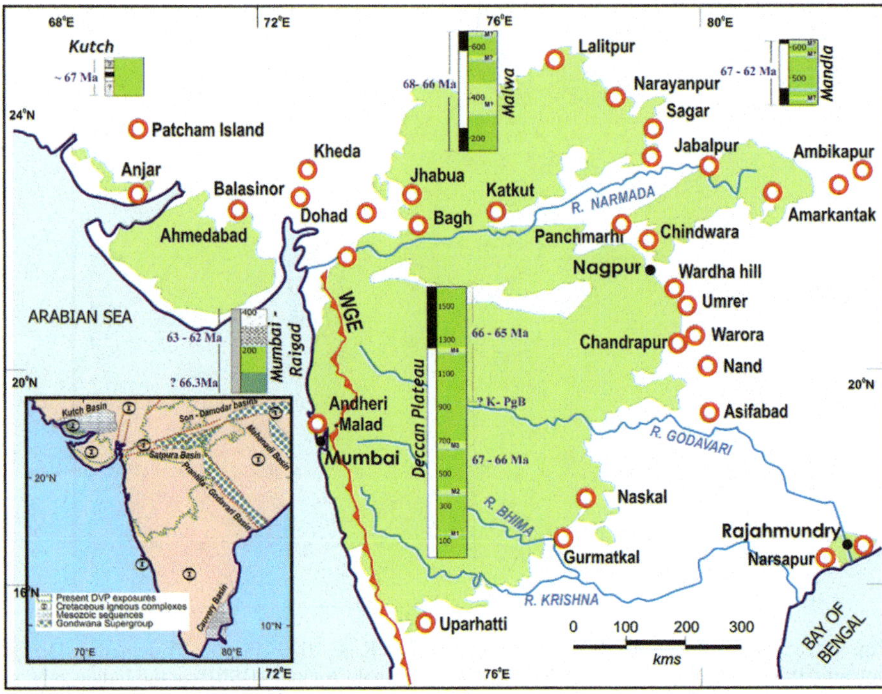

Fig. 5 Fossil localities of Late Cretaceous fossils (marked with red open circles) associated with the DVP. Lithologs, paleomagnetic logs and the ages of the lava flows from different subprovinces (modified after Kale et al. 2020b; with geochronological inputs from Schoene et al. 2021; Kale and Pande 2022; Tholt et al. 2023) are given for reference. Inset depicts the larger sedimentary basins that evolved within the Indian Peninsula prior to the eruption of the Deccan lavas

~20°S to its present location: see Fig. 4) across the equator, the Indian plate was moving through the tropical region. It enjoyed a warm tropical climate throughout this period and supported a rich assemblage of terrestrial flora and fauna, notably dinosaurian fossils and egg-clutches (Kapur and Khosla 2018). The Late Cretaceous saw the rifting and separation of Madagascar from India during its passage over the Réunion hotspot, leading to the deposition of the Upper Gondwana and Bagh—Lameta sediments along the (present day course of) River Narmada; including a speculated marine incursion (Kumari et al. 2020; Shitole et al. 2021). Large components of these biotic assemblages perished during the terminal-Cretaceous mass extinction event.

The sequences that host them are exposed all around the fringes of the DVP (Fig. 5) and provide exciting insights into the pre-eruptive ecology and environments that existed on the Indian subcontinent. Their study provides direct clues to the impact of volcanism on the sedimentary basins, exposed continental surfaces and contemporary paleoecology. It is noteworthy that one such location in Dhar district is being promoted as a geopark (Shitole et al. 2023) and there is scope for identifying more such locations as the scientific community turns its attention to the subject.

4 Cretaceous–Paleogene Boundary and Environmental Impact

Geochronological and paleomagnetic studies in the DVP in the last couple of decades (Sprain et al. 2019; Eddy et al. 2020; Schoene et al. 2021, 2022; Kale and Pande 2022; Tholt et al. 2023) have shown that a bulk (>85%) of the Deccan volcanism occurred in a short span of time of about 0.8–0.9 million years beginning during the C30n-C29r reversal and waned shortly after the C29r-C29n reversal in at least 2 major pulses. It is very likely that the two pulses occurred on either side of the Cretaceous–Paleogene (K-Pg) Boundary (dated @ 66.052–66.030 Ma; Renne et al. 2013; Sprain et al. 2018). This coincidence of the Deccan volcanism straddling the K-Pg Boundary that witnessed a dramatic faunal turnover highlighted by extinction of the dinosaurs and emergence of mammalian fauna in the terrestrial realm has drawn attention of the global scientific community to it. Sporadic third phase of Deccan volcanism (~62 ± 1 Ma: Kale et al. 2020b; Mittal et al. 2021) is only exposed along the edge of western DVP but may have had a wider imprint that is now downfaulted along the western coastal margin of the Indian Peninsula and submerged below the Arabian Sea (Bhattacharya and Yatheesh 2015; Nair et al. 2021; Dole et al. 2022).

A comparison of the erupted volumes (several cubic kilometers per year for more than a million years) from the DVP with some recent volcanic eruptions gives the perception of the impact that such CFPBs have on the local and global environment. The destruction of the rich city of Pompeii caused by the eruption of Mount Vesuvius (79 A.D.) is one of the best-known volcanic catastrophes in urban history. The 1783–1784 Laki eruptions in Iceland emitted approximately 120 megatons of SO_2 to the atmosphere and the acidic aerosol veil covered the northern hemisphere for not less than 5 months. The eruption of Mount St Helens in March 1980 produced a blast zone with a radius of almost 500 km and ejected superheated ash and stone across an area of more than 25 km. The disruptions of air traffic by the gaseous clouds from eruptions of Grimsvőtn (Fig. 6b, c), Eyjafjallajokull and Katla from Iceland during the 2010–2018 period are well known.

None of these volcanoes produced lava volumes exceeding 10 km^3 whereas each eruptive event in the Deccan volcanism erupted 1000s of km^3 of lava. On this background, one can only imagine the damage and environmental crisis caused by the Deccan flood basalts that are more than ten thousand times larger than any of these volcanic activities (Self et al. 2014). Several mass extinction events in geological history have been linked to CFBPs (Ernst and Youbi 2017). The presence of distinct Iridium anomalies and suspected impact ejecta in all the K-Pg boundary sequences across the world led to the theory of an extraterrestrial cause of this mass extinction (e.g. Richards et al. 2015). The growing evidence of the temporal linkage of the Deccan volcanism to the K-Pg Boundary and its environmental impact (Calegaro et al. 2023) leading to the faunal turnover across the K-Pg boundary is undeniable.

5 Concluding Remarks

The above compilation provides a snapshot of only some of the features of the DVP that have emerged in the last few years, leading to revision of several traditional concepts. The geoheritage significance of the DVP as a region is borne out by the facts that (a) hosts some of the largest volcanic eruptions to have ever occurred in the Earth's history, (b) is arguably the primary cause of the extinction of dinosaurs and many other biotic forms and (c) is a host to some of the exquisite exposures of volcanic features and for the study of flood basalt volcanism.

Their erosion in the ensuing 65 million years during the northward flight of the Indian Plate has yielded a picturesque landscape of cascading waterfalls, steep escarpments, dense tropical forests, wild-life sanctuaries (including at least six tiger reserves) and a large variety of geomorphic features that are popular tourist spots of international and national repute like Mahabaleshwar, Matheran, Igatpuri, Saputara, Amarkantak, Choral valley and Patalpani to name a few. What also need to be recognised from the geoheritage perspective are some unique geomorphic features such as volcanic hills (Fig. 6a) which have heritage forts atop them, entrenched meanders (Fig. 6b), pot-holed riverbeds (Fig. 6c) and natural arches (Fig. 6d) that provide concrete evidence of the erosional patterns, river basin evolutions, past climates, and tectonic and geomorphic heritage of the DVP (Dole et al. 2022). Recognition of such features and their heritage significance awaits documentation.

The caves carved out of the Deccan basalts at Ajanta, Ellora and Elephanta are already recognised as UNESCO World Heritage Sites. Numerous smaller rock-caves with carvings (e.g. Karla (near Pune), Pandav Leni, near Nasik) and temple structures have a historic and aesthetic value and are already known as tourist destinations. They provide ready-to-study walls and multidimensional artificial cuttings where numerous volcanological features of the basaltic lavas can be studied. They need additional endorsement as geoheritage sites (Sheth et al. 2017; Sheth 2023).

Fig. 6 (**a**) Tiger point (18°41′39″N, 73°22′46″E) near Lonavala, Maharashtra. A unique example of an isolated conical hill towering (> 600 m high) towering over the Konkan coastal plains. (**b**) Entrenched meander channel of Mandhol River indicating enhanced incision by antecedent drainage in an uplifted terrain from western DVP (Kale et al. 2017). (**c**) Pothole-driven incision by the Ghod river (near Nighoj) in a mature river valley indicating rejuvenation of a river due to neotectonic movements of the host terrain (see Kale et al. 2017). (**d**) One of the largest standing natural arch in a flood basalt province is located in the Ane Ghat section of Ahmednagar district, Maharashtra (see Kale et al. 2017)

Fig. 6 (continued)

Fig. 6 (continued)

The Lonar crater, arguably the only known impact crater in a basaltic terrain in the world (Bodas and Sen 2014) is located within the DVP. Several other lesser recognized, but spectacular geomorphic curiosities and volcanic features can become geotourism hotspots and qualify as Global Geoparks. Taken together, the strong case for the Deccan Volcanic Province as a geoheritage region and places within it at geoheritage sites awaits formal documentation, recording and recognition.

References

Basu AR, Chakrabarty P, Szymanowski D, Iganez-Mejia M, Schoene B, Ghosh N, Bastian Georg R (2020) Widespread silicic and alkaline magmatism synchornous with the Deccan trap flood basalts, India. Earth Planetary Sci Lett 552(116616) 10 p. https://doi.org/10.1016/j.epsl.2020.116616

Bhattacharya GC, Yatheesh V (2015) Plate-tectonic evolution of the deep ocean basins adjoining the Western Continental Margin of India—a proposed model for the early opening scenario. In: Mukherjee S (ed) Petroleum geosciences: Indian contexts, 61 p. https://doi.org/10.1007/978-3-319-03119-4_1

Bodas MS, Sen B (2014) The Lonar crater: the bast preserved impact crater in the basaltic terrain. In: Kale VS (ed) Landscapes and landforms of India. Springer Science, pp 223–230

Bryan SE, Ernst RE (2008) Revised definition of large igneous provinces (LIPs). Earth Sci Rev 86:175–202

Calegaro S, Baker DR, Renne PR, Melluso L, Geraki K, Whitehouse MJ, de Min A, Marzolli A (2023) Recurring volcanic winters during the latest cretaceous: sulfur and fluorine budeges of Deccan trap lavas. Sci Adv 9., 11 p. https://doi.org/10.1126/scieadv.adg8284

Coffin MF, Pringle MS, Duncan RA, Gladczenko TP, Storey M, Muller RD, Gahagan LA (2002) Kerguelen hotspot magma output since 130 Ma. J Petrol 43:1121–1139

Dietterich HR, Lev E, Chen J, Richardson JA, Cashman KV (2017) Benchmarking computational fluid dynamics models of lava flow simulation for hazard assessment, forecasting and risk management. J Appl Volcanol 6. https://doi.org/10.1186/s13617-017-0061-x

Dole G, Das S, Kale VS (2022) Tectonic framework of geomorphic evolution of the Deccan Volcanic Province, India. Earth-Sci Rev 228.: 36 p. https://doi.org/10.1016/j.earscirev.2022.103988

Duraiswami RA, Bondre NR, Managave S (2008) Morphology of rubbly pahoehoe (simple) flows from the Deccan Volcanic Province: implications on style of emplacement. J Volcanol Geothermal Res 177:822–836. https://doi.org/10.1016/j.jvolgeores.2008.01.048

Eddy MP, Schoene B, Samperton KM, Keller G, Adatte T, Khadri SFR (2020) U-Pb zircon age constraints on the earliest eruptions of Deccan large Igneous Province, Malwa Plateau, India. Earth Planetary Sci Lett 540(116249) 9 p. https://doi.org/10.1016/j.epsl.2020.116249

Ernst RE, Youbi N (2017) How large igneous provinces affect global climate, sometimes cause mass extinctions and represent natural markers in the geological record. Palaeogeogr Palaeoclimatol Palaeoecol 478:30–52

Hyman DMR, Diettrerich HR, Patrick MR (2022) Toward next-generation lava flow forecasting: development of a fast, physic-based lava propagation model. J Geophys Res – Solid Earth 127. https://doi.org/10.1029/2022JB024998

Jay AE, Widdowson M (2008) Stratigraphy, structure and volcanology of the SE Deccan continental flood basalt province: implications for eruptive extent and volumes. J Geol Soc Lond 165:177–188. https://doi.org/10.1144/0016-76492006-062

Joshi N, Gadgil V (2024) Columnar trap dyke of Potla, district Dewas, Madhya Pradesh, India: a geological marvel and potential geotourism destination. Geoheritage 16., 14 p. https://doi.org/10.1007/s12371-023000907-9

Kale VS (2020) Cretaceous volcanism in Peninsular India: Rajmahal—Sylhet & Deccan Traps. In: Gupta N, Tandon SK (eds) Geodynamics of the Indian Plate. Springer, pp 233–289. https://doi.org/10.1007/978-3-030-15989-4_8

Kale VS, Pande K (2022) Reappraisal of duration and eruptive rates in Deccan Volcanic Province, India. J Geol Soc India 98:7–17. https://doi.org/10.1007/s12594-022-1921-5

Kale VS, Dole G, Upasani D, Patil Pillai S (2017) Deccan Plateau uplift: insights from parts of Western Uplands, Maharashtra, India. In Mukherjee S, Misra AA, Calves G, Necok M (eds) Geological Society London Special Publication, vol 445, pp 11–46. https://doi.org/10.1144/SP445.2

Kale VS, Bodas MS, Chatterjee P, Pande K (2020a) Emplacement history and evolution of the Deccan Volcanic Province, India. Episodes 43:278–299. https://doi.org/10.18814/epiiugs/2020/20015

Kale VS, Dole G, Shandilya P, Pande K (2020b) Stratigraphy and correlations in Deccan Volcanic Province, India: quo vadis? Geol Soc Am Bull 132:588–607. https://doi.org/10.1130/B35018.1

Kale VS, Dole G, Patil Pillai S, Chatterjee P, Bodas MS (2021) Morphological types in the Deccan Volcanic Province, India: implications for emplacement dynamics of continental flood basalts. In Srivastava RK, Ernst RE, Buchan KL, de Kock M (eds) Geological Society London Special Publication, vol 518, pp 341–396. https://doi.org/10.1144/SP518-2020-246

Kapur VV, Khosla A (2018) Faunal elements from the Deccan volcano-sedimentary sequences of India: a reappraisal of biostratigraphic, palaeoecologic, and palaeobiogeographic aspects. Geol J 32p:2797–2828. https://doi.org/10.1002/gj.3379

Keszthelyi L, Self S (1998) Some physical requirements for the emplacement of long basaltic flows. J Geophys Res 103:27447–27464. https://doi.org/10.1029/98JB00606

Keszthelyi L, Self S, Thordarson T (1999) Application of recent studies on the emplacement of basaltic lava flows to the Deccan Traps. In: Subbarao KV (ed) Deccan Volcanic Province. Geological Society India Memoirs, vol 43, pp 485–520

Krishnamurthy P (2020) The Deccan Volcanic Province (DVP), India: a review. Part 1. J Geol Soc India 96:9–35. https://doi.org/10.1007/s12594-020-1501-5

Kumari V, Tandon SK, Kumar N, Ghatak A (2020) Epicontinental Permian—cretaceous seaways in Central India: the debate for the Narmada versus Godavari rifts for the cretaceous—tertiary incursion. Earth Sci Rev 211., 24 p. https://doi.org/10.1016/j.earscirev.2020.103284

Mittal T, Richards MA, Fendley IM (2021) The magmatic architecture of continental flood basalts I: observations from the Deccan traps. J Geophys Res Solid Earth 126., 54 p. https://doi.org/10.1029/2021JB021808

Nair N, Pandey DK, Pandey A, Prerna R (2021) Seismic stratigraphy and the sedimentation history in the Laxmi Basin of the eastern Arabian Sea: constraints from IODP expedition 355. Geosci Front 12(101111) 14 p. https://doi.org/10.1016/j.gsf.2020.11.008

Patil Pillai S, Pattabhiram K, Dole G, Shandilya P, Upasani D, Pande K, Kale VS (2022) Geometry and age of a mafic dyke emplaced along the Bhetkheda—Mohana lineament, Central Narmada valley, Deccan Volcanic Province. J Earth Syst Sci 131(121):17. https://doi.org/10.1007/s12040-022-01846-9

Renne PR, Deino AL, Hilgen FJ, Kuiper KF, Mark DF, Mitchell WS III, Morgan LE, Mundil R, Smit J (2013) Time scales of critical events around the cretaceous-paleogene boundary. Science 339:684–687

Richards MA, Alvarez W, Self S, Karlstrom L, Renne PR, Manga M, Sprain CJ, Smit J, Vanderkluysen L, Gibson SA (2015) Triggering of the largest Deccan eruptions by the Chicxulub impact. Geol Soc Am Bull 127:1507–1520

Sangode SJ, Dongre A, Bhagat A, Meshram D (2022) Paleomagnetic inclination anomaly in the Deccan traps and its geodynamic implications over the Indian plate. J Earth Syst Sci 131(178):12. https://doi.org/10.1007/s12040-022-001917-x

Schoene B, Eddy MP, Samperton KM, Keller CB, Keller G, Adatte T, Khadri SFR (2019) U-Pb constraints on pulsed eruption of the Deccan traps across the end-cretaceous mass extinction. Science 363:862–866. https://doi.org/10.1126/science.aau2422

Schoene B, Eddy MP, Keller CB, Samperton KM (2021) An evaluation of Deccan traps eruptions rates using geochronologic data. Geochronology 3:181–198

Self S, Keszthelyi L, Thordarson T (1998) The importance of Pahoehoe. Annu Rev Earth Planetary Sci 26:81–110. https://doi.org/10.1146/annurev.earth.26.1.81

Self S, Schmidt A, Mather TA (2014) Emplacement characteristics, time scales, and volatile release rates of continental flood basalt eruptions on Earth. In: Keller G, Kerr A (eds) Volcanism, impacts and mass extinctions: causes and effects. Geological Society America, Special Papers, vol 505, pp 319–337. https://doi.org/10.1130/2014.2505(16)

Self S, Mittal T, Jay AE (2021) Thickness characteristics of pahoehoe lavas in the Deccan Province Western Ghats, India and in continental flood basalt provinces elsewhere. Front Earth Sci 22p. https://doi.org/10.3389/feart.2020.630604

Shandilya P, Chatterjee P, Pattabhiram K, Bodas M, Pande K, Kale VS (2020) Rajgad GPB: a mega-porphyritic flow field, western Deccan Volcanic Province, India. J Earth Syst Sci 129(113):11. https://doi.org/10.1007/s12040-020-1375-3

Sheth H (2023) The volcanic geoheritage of the Ajanta and Ellora caves, Central Deccan traps, India. Geoheritage 15(39):22p. https://doi.org/10.1007/s12371-023-00809-w

Sheth H, Samant H, Patel V, D'Souza J (2017) The volcanic geoheritage of the Elephanta caves, Deccan Traps, Western India. v.9:359–372. https://doi.org/10.1007/s12371-016-0214-z

Shitole AD, Patil SJ, Damgawn JL, Joseph JK (2021) Amended lithostratigraphy of the cretaceous Bagh group, western lower Narmada Valley, India: a comparison with pervasive Tethyan basins. J Geol 56:5058–5093. https://doi.org/10.1002/gj.4224

Shitole AD, Gupta V, Gidwani L, Verma V, Rajora S, Anand S, Tripathi SC (2023) Bagh dinosaur national part regions, Dhar district, Madhya Pradesh: a potential UNESCO global geopark. Geoheritage 15:116. https://doi.org/10.1007/s12371-023-00881-2

Sprain CJ, Renne PR, Clemens WA, Wilson GP (2018) Calibration of chron c29r: new high-precision geochronologic and paleomagnetic constraints from the hell creek region, Montana. Bull Geol Soc Am Bull 130:1615–1644

Sprain CJ, Renne PR, Vanderkluysen L, Pande K, Self S, Mittal T (2019) The eruptive tempo of Deccan volcanism in relation to the cretaceous paleogene boundary. Science 363:866–870

Tarquini S, de'Michieli Vitturi M, Jensen EH, Pedersen GBM, Barsotti S, Coppola D, Pfeffer MA (2019) Modelling lava flow propogation over a flat landscape using MrLavaLoba: the case of the 2014–2015 eruption at Holuhraun, Iceland. Ann Geophys 62., 18 p. https://doi.org/10.4401/ag-7812

Tholt AJ, Renne PR, Marzoli A, Vanderkluysen L, Mahobey D, Samant B, Dhobale A, Pande K, Self S (2023) Geochronological constraints on the evolution and petrogenesis of the Malwa plateau subprovince of the Deccan traps. Geochem Geophys Geosyst 24., 15 p. https://doi.org/10.1029/2023GC.011137

Valdiya KS (2016) The making of India: geodynamic evolution, 2nd edn. Springer. 924 p

Walker GPL (2009) The endogenous growth of pahoehoe lava lobes and morphology of lava-rise edges. In: Thordarson T, Self S, Larsen G, Rowland SK, Höskuldsson A (eds) Studies in volcanology: the legacy of George Walker. IAVCEI/Geological Society, London, Special Publications, vol 2, pp 17–32

West WD (1959) The source of the Deccan trap flows. J Geol Soc India 1:44–52

Zhang Z, Li G, He X, Zhang L, Dong S, Qing C, Liang W, Han S (2023) The evolution of Kerguelen mantle plume and breakup of eastern Gondwana: new insights from multistage cretaceous magmatism in the Tethyan Himalaya. Gondwana Res 119:68–85. https://doi.org/10.1016/j.gr.2023.03.009

When Peninsular India Experienced First Marine Incursion ~290 Ma Ago

Shivranjan Kumar Bharti

Abstract During early Permian time, dramatic climatic and palaeogeographic changes took place in the Earth's evolutionary history. These phenomenons have been preserved globally in the wide apart localities at the base of Gondwana sequences. Among these, the prime locations of Indian peninsula are Dudhinala Section in Jharkhand and Manendragarh in Chhattisgarh exhibit signatures of marine incursions during 290 Ma ago.

Keywords Marine Gondwana · Early Permian · Manendragarh · Dudhinala

1 Introduction and Geoscientific Context

India along with the continents of southern hemisphere was a part of supercontinent Gondwanaland (Suess 1885), that existed as a single landmass since Cambrian till its eventual break up in phases during Jurassic-Lower Cretaceous. India, Madagascar, Western and Northern Australia, East Antarctica formed East Gondwana while Africa and South America were part of West Gondwana. There are multiple striking geological events that have happened on the Earth's surface during ~290 ± 10 Ma ago and which draw the geologist's attention. Permian Period (299–252 Ma) is very significant for the Indian subcontinent, because 99% of the Indian coal resource has been formed during this time interval (Mukhopadhyay et al. 2010). Moreover, the Early Permian period was an exceptional phase in the Earth's history when the precursors of the modern continents were assembled in the form of two big landmasses (Gondwana and Laurasia; Scotese 1997) and the remarkable feature of Permian biogeography 'The Tethys' came into existence (Fig. 1). It was very extensive and stretched from the Pyrenees in Spain and the Atlas Mountains in north west Africa, through the Alpine region, south east Europe, Turkey, Syria and Iran into Baluchistan,

S. K. Bharti (✉)
Geological Survey of India, State Unit-Jharkhand, Ranchi, Jharkhand, India

S. C. Tripathi et al. (eds.), *Geoconservation and Geotourism Potential of India*, Society of Earth Scientists Series, https://doi.org/10.1007/978-3-031-81017-6_10

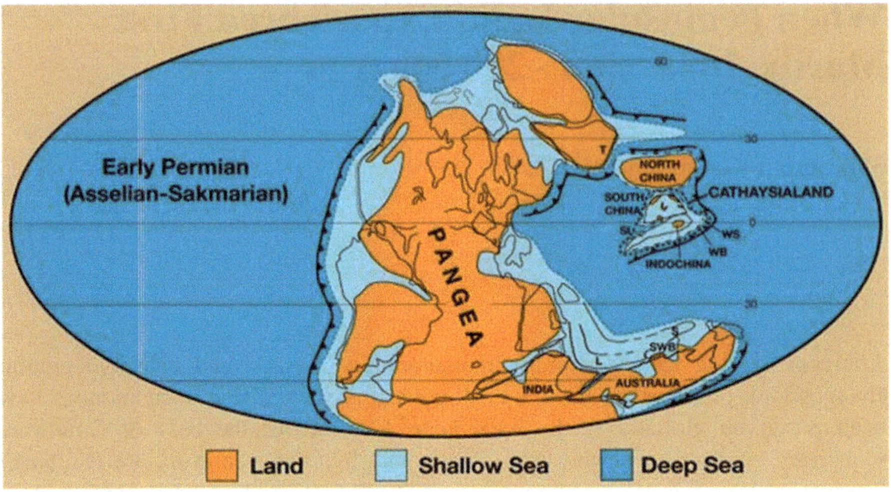

Fig. 1 Early Permian biogeography (Source: www.britishpaleoclimateinthepalaeoz.weebly.com)

Salt Range in the Pakistan, the Himalayas, the Assam-Arakan ranges and further on into the East Indian Archipelago. The Early Permian (Sakmarian) was a time of rapid global climatic change as the major glacial period came to an end (Crowell 1995). However, vast ice sheets covered the polar region of Gondwana and extended to mid-paleolatitudes (Ziegler et al. 1997) and there is evidence for continental ice sheets covering a total area of 17.9–22.6 × 106 km^2 (Isbell et al. 2003). The sedimentary rocks comprised of diamictite, conglomerate, sandstone, shale, rhythmically alternating shale and sandstone with or without dropstones and thin marl bands were deposited in this major glacial episode occur in various types of sedimentary basins—cratonic basins (e.g. Kalahari, Rubh al Khali and Paraná basins); rift basins (e.g. Damodar, Son, Mahanadi basins); foreland basins (e.g. Sydney and Karoo basins).

In the middle of twentieth century, discovery of the marine fossils from the central India viz. Umaria and Manendragarh changed the notion of entirely fresh water deposits of Gondwanas of peninsular India. Till then, the Indian peninsula was never considered to have been invaded by seas after Precambrian and before Jurassic or Cretaceous Periods and even then marine incursions were known mostly along the Palaeo-Tethyan margin (stretching from Kashmir–Spiti-Garhwal–Sikkim Himalaya). Thus, the discovery of marine fossils from central India was of great significance, which provides a pin-hole vision to connect the paleobiogeographic thread among the different landmass and paleoclimatic insight into the glacial-deglacial phases. It's became momentous, as these fossils have been preserved in the rock successions from where the supercontinent 'GONDWANALAND' has gained its name, that is based on the 'GOND' tribe of Madhya Pradesh. This fossiliferous bed along with the glacial tillites forms the oldest unit of the approximately 5 km thick strata, deposited over 200 million years, from Upper Carboniferous to Lower Cretaceous (Veevers and Tewari 1995), which are preserved in rivers basins Damodar, Son, Mahanadi, Godavari etc. or the linear hill ranges like Satpura and Rajmahal and are clubbed into Gondwana Supergroup.

The faunal assemblage reported at Manendragarh and Umaria represents the marine organisms living in the sea that transgressed into the Indian Peninsula as the ice sheets melted. A *Eurydesma*-bearing faunal assemblage of Lower Permian (Sakmarian) age within the Lower Gondwana (Talchir Formation) is closely related to the Lower Permian marine fauna found at the base of Gondwana sequence in other parts of the Gondwanaland, viz., South Africa, South America, India, Australia and Antarctica. The remnants of this transgression are found as small and isolated outcrops in the wide apart localities near Subansiri in Arunachal Pradesh, Khemgaon in Sikkim, Rajhara (Daltonganj) in Jharkhand, Manendragarh in Chhattisgarh, Umaria in Madhya Pradesh, and near Bap and Badhura in Rajasthan. The fossil assemblage of Daltonganj, Manendragarh are rich in *Eurydesma* fauna similar to Eastern Himalaya; while that of Umaria and Salt Range are rich in *Productus*; which lead Shastry and Shah (1964) to visualise two marine incursions, one from the east i.e. from Khemgaon to Manendragarh during the Sakmarian time (~293 Ma Ago) and other from the west i.e. from the Salt Range to the Umaria during Artinskian Time (~290 Ma Ago). However, subsequently, discovery of cold water *Eurydesma* fauna from the Shahpur, Betul, which is situated between the way of Salt Range-Bap/Badhaura–Umaria incursion, where a productid fauna was expected that could not be found, led the Ghosh (2003) to propose only one marine incursion which took place along the rift valley (Son-Narmada-Damodar lineament) from the Khemgaon in the east and reached upto the Shahpur, Betul M.P via Dudhi Nala and Rajhara (Daltonganj) in Jharkhand, Manendragarh in Chhattisgarh (Fig. 2).

There are reports of early Permian marine fossil from the Athgarh and Palar Basin which shows the prevalence of marine condition in the south-eastern coastal areas of peninsular India. Amongst the marine fauna, a cold water fauna consisting of bivalves such as *Eurydesma* and *Deltopecten*, and brachiopods such as *Lyonia* and *Trigonotreta*, were established in the earliest post-glacial marine transgressions. Above this is a more diverse, increasingly warmer, temperate fauna, including brachiopods, bryozoans, bivalves, gastropods, conularids, small foraminifers, asterozoans, blastoids and crinoids. The paleontological data from the other parts of Gondwanaland too indicate that the synchronous deposition during the post-glacial phase and are in consistent with the farfield evidences. For example, Early Permian farfield isotope studies from brachiopods from the southern Urals, show a $\delta_{18}O$ decline of 2.5‰ in the Early Permian (Asselian to Artinskian) and stable $\delta_{13}C$ values of around +4.3‰ in the same period. Which is most likely cause for the decline in $\delta_{18}O$ is the return of isotopically light waters to the oceans from melting of glaciers at high latitudes (Korte et al. 2005).

Initial sedimentation in all the basins of Gondwanaland took place due to melting of glaciers. Deglaciation caused rise in sea level and during high stands, seawater entered into the inland basins from the main Tethys, which is manifested by the occurrence of Eurydesma and other marine fauna all over the Gondwanaland countries during Early Sakmarian at around 290 + 4 Ma (Berthold et al. 1999). Following

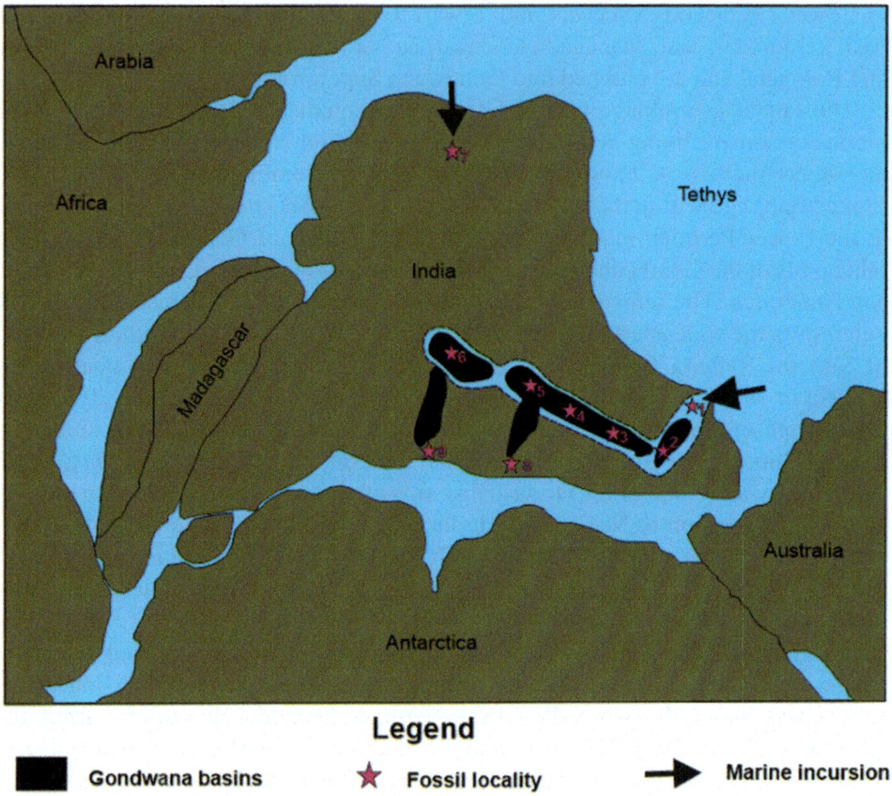

Fig. 2 Inferred paleobiogeographic map of Peninsular India during early Permian time (modified after, Ghosh 2003; Mukhopadhyay et al. 2010). Fossil localities—1. Khemgaon, Sikkim; 2. Dudhinala, west Bokaro basin; 3. Daltonganj, Jharkhand; 4. Manendragarh; Chattisgarh; 5. Umaria, M.P.; 6. Shahpur, Betul, 7. Bap & Badhaura, Rajasthan, 8. Athgarh, Orissa, 9. Palar, Godawari

the Sakmarian marine transgression, near shore environment prevailed in many basins in the interior parts of the Gondwanaland with the advent of fluvial environment in the marginal part of the basins under periglacial climate (Chakraborty and Ghosh 2003; Eyles et al. 2003). The lower part of Barakar Formation in India can be equated with these deposits. These entire aforementioned phenomenons are excellently archived as sedimentary structures and fossiliferous horizon in Dudhinala Section, Hazaribagh, Jharkhand and in Manendragarh, Chhattisgarh.

Fig. 3 Fossiliferous glaciated boulder bed, Manendragarh

2 Archives of Marine Incursions

Site 1

Description: Marine Gondwana Fossil Park at Manendragarh, Sarguja district, Chattisgarh is an unique exposure of fossiliferous marine Permian (280–240 Ma) rocks of Talchir Formation belonging to Gondwana Supergroup. It is exposed for a length of about 1 km upstream to the confluence of Hasdeo River and Hasia nala. The marine fauna is represented by the dominance of pelecypods/lamellibranchs like *Eurydesma* and *Aviculopecten* within shale, besides Bryozoans, Crinoids and foraminifers (Figs. 3 and 4).

Location: The area can be approached by road from Nagpur via Jabalpur (NH-7), Katni, Shahdol, Burhar, Anupur and Kotma (SH-14). The area is located on the right bank of Hasdeo River, near the railway bridge between milestone 937 and 938 of the Anuppur-Chirmiri branch of Southeastern Railway and about 2.5 km southeast of Manendragarh railway station near Ammakherwa village. The fossil park can be visited round the year except rainy season, when the fossiliferous horizon is flooded by Hasdeo river.

Site 2

Description: Dudhi Nala exhibit a thick pile of glacial and periglacial deposits of Talchir Formation. The prominent sedimentary features, as noted in the section, include tillites (sediment deposited by glacier), faceted pebble, dropstones and different mass flow conglomerates of glaciogenic origin. These glacial deposits is overlain by sandstones which contains shore face features like wave ripples

Fig. 4 *Aviculopecten mitchelli,* Manendragarh

Fig. 5 (**a** and **b**) Glacial imprints in Talchir Formation, Dudhi Nala Section

associated with climbing ripples, hummocky cross-stratifications etc. These glacio-marine signatures occur together one after other fascinates to perceive the paleo-environmental fluctuation. Other associated sedimentary structures include penecontemporaneous deformation features/soft sedimentary deformation structure (SSD) like load and flame structures, convolute laminae, slump structures etc. (Figs. 5, 6, 7 and 8) reveal the occurrence of paleoseismic phenomenon during the

Fig. 6 (**a**) Dropstone within siltstone; (**b**) Dropstone within sandstone, Dudhi Nala Section

Fig. 7 (**a**) Flame structure; (**b** and **c**) Convolute lamination, Dudhi Nala Section

deposition of glacio-marine sediments in the Dudhinala and which have also been observed in other Gondwana Basins. Towards east, the Karharbari Formation gradationally overlies the Talchir Formation or overlaps it to rest on basement. In spite of having every potentiality to be declared as a geohertitage site, Dudhinala has been neglected.

Location: Dudhi Nala is located partly in Hazaribagh and Ramgarh districts of Jharkhand, falling in Survey of India Toposheet Nos. 73 E/5; latitude: 23°74′N, longitude: 85°49′E. Ranchi, the state capital is situated about 70 km towards SSW and Hazaribagh town is situated about 40 km NNW of this area. Ranchi is well

Fig. 8 (**a**) Wave ripple; (**b**) wave generated structure, Dudhi Nala Section

Fig. 9 "Gondwana Marine Fossil Park" promoted by State forest department, Chhattisgarh

connected by flights and railways with major cities of the country. Ramgarh is the nearest town at about 30 km from field area is a convenient option to stay. To reach there one can take two roads to this place. The first is through a newly laid expressway that traverses the scenic Patratu valley and verdant hills. The other route is the on the more austere National Highway-33 that goes via Ramgarh cantonment.

It is very unfortunate that due to the construction of double railway line, the famous Umaria Marine Bed has been totally vanished and the Bap-Badhaura fossiliferous bed is being used as construction material by the local people. However, with the effort of Geological Survey of India the fossil locality of Manendragarh, on the bank of Hasdeo River has been declared as National Geological Monument and a "Gondwana Marine Fossil Park" is developed and maintained by State forest department as a small tourist place. However, this is a geodiversity hotspot and does not belong to biodiversity need to be understood by authorities (Fig. 9).

References

Berthold B, Harald S, Volker L, Richard A (1999) The geochronology and significance of ash-fall tuffs in the glaciogenic carboniferous-Permian Dwyka Group of Namibia and South Africa. J Afr Earth Sci 29:33–49

Chakraborty C, Ghosh SK (2003) Depositional record of tidal-flat sedimentation in the Permian coal measures of Central India: Barakar formation, Mohpani coalfield, Satpura Gondwana Basin. Gondwana Res 6:817–827

Crowell JC (1995) The ending of the Late Paleozoic ice age during the Permian paper. In: Scholle PA, Peryt TM, Ulmer-Scholle DS (eds) The Permian of northern Pangea. Palaeogeography, Palaeoclimates, Stratigraphy. Springer, pp 62–74

Eyles CH, Mory AJ, Eyles N (2003) Carboniferous-Permian facies and tectono-stratigraphic successions of the glacially influenced and rifted Carnarvon basin, Western Australia. Sediment Geol 155:63–86

Ghosh SK (2003) First record of marine bivalves from the Talchir formation of the Satpura Gondwana basin, India: Palaeobiogeographic implications. Gondwana Res 6:312–320

Isbell JL, Miller MF, Wolfe KL, Lenaker PA (2003) Timing of late Paleozoic glaciation in Gondwana: was glaciation responsible for the development of northern hemisphere cyclothems? Geol Soc Am Spec Pap 370:5–24

Korte C, Jasper T, Kozur HW, Veizer J (2005) $\delta^{18}O$ and $\delta^{13}C$ of Permian brachiopods: a record of seawater evolution and continental glaciations. Palaeogeogr Palaeoclimatol Palaeoecol 224:333–351

Mukhopadhyay G, Mukhopadhyay SK, Manas R, Parui PK (2010) Stratigraphic correlation between different Gondwana basins of India. Geol Soc India 76:252–266

Scotese CR (1997) Paleogeographic Atlas, PALEOMAP Progress report 90–0497, Department of Geology, University of Texas at Arlington, Arlington, TX, pp 45–48

Shastry MVA, Shah SC (1964) Permian marine transgression in Peninsular India; 22nd International Geological Congress proceedings, pt 19, pp 139–150

Suess E (1885) Das Satilitz der Ende, Brand 1, Wien, Leipzig

Veevers JJ, Tewari RC (1995) Gondwana master basin of peninsular India between Tethys and the interior of the Gondwanaland province of Pangea. Geol Soc Am Memoir 187

Ziegler AM, Hulver ML, Rowley DB (1997) Permian world topography and climate. In: Martini IP (ed) Late glacial and post-glacial environmental changes-quaternary, carboniferous- permian and proterozoic. Oxford University Press, New York, pp 111–146

Coprolites: Geological Signatures of Dietary Habits of Pre-historic Indian Vertebrates

Vivesh V. Kapur

Abstract Fossilized faecal matter (coprolite) of pre-historic vertebrates provide interesting records of the dietary habit(s), change(s) in the dietary pattern (if any), and assist to reconstruct the surrounding habitat(s) in the geological past. In India, the research on coprolites was initiated in the later half of the eighteenth century and has since focussed on the large (>10 cm in length) Maastrichtian Type-A coprolites from the infratrappean deposit of Pisdura, Maharashtra State, central India. However, in the past few years new data on coprolites from co-eval and younger (Paleogene-Neogene) sedimentary sequences has been recorded. Overall, coprolite records are a rarity in India as gaps exist in the fossil record of these ichnofossils. These ichnofossils occur sporadically in geographically distant locations, but have the potential to reveal crucial information on pre-historic animal-environment interactions. In view of the palaeontological potential of coprolite ichnofossils and popularity of ongoing geoconservation efforts in India, it is necessary that existing and newly discovered fossiliferous (coprolite-yielding) sites and/or regions should be sustainably protected through standard policies following an inclusive approach with local communities, state, and central agencies.

Keywords Coprolite ichnofossils · Geoconservation · Geoheritage · Infra- and intertrappean · Paleogene-Neogene

V. V. Kapur (✉)
Vertebrate Palaeontology and Preparation Laboratory, Birbal Sahni Institute of Palaeosciences, Lucknow, India
e-mail: viveshvir_kapur@bsip.res.in

S. C. Tripathi et al. (eds.), *Geoconservation and Geotourism Potential of India*, Society of Earth Scientists Series, https://doi.org/10.1007/978-3-031-81017-6_11

1 Introduction

"*You are what you eat*" Lindlahr (1942) is a phrase construed in line with the idea of linking dietary habit(s) and health amongst humans in modern times. The statement holds true for other animal groups as well, and quite possibly one of the many reasons that generated curiosity in our minds to know the diet of animals (e.g. dinosaurs) that thrived on our planet in the geological past. Coprolites (fossilized faecal matter) are the geological signatures of ancient animals that are recorded in millions of years to a few thousand-year-old rocks. They preserve direct evidence of the diet consumed by the producer animal and thus provide an opportunity to palaeontologists to understand the dietary habit(s) in extinct animal groups, and the ecosystem around these animals (Qvarnström et al. 2024 and references therein). Fortunately, a wealth of these ichnofossils (containing partly digested and/or undigested food remains) is known to be preserved in the rock records of India (Khosla et al. 2015 and references therein). However, the coprolite research in India has focussed on the late Cretaceous (Maastrichtian) Type-A morphotype from the Lameta Formation (infratrappean) deposit at Pisdura, central India (Matley 1939a; Prasad et al. 2005; Khosla et al. 2015). Kapur et al. (2006) recorded vertebrate coprolites from the Lotkheri intertrappean deposit, Mandsaur District, Madhya Pradesh, central India. However, only recently additional coprolites from Lotkheri have been studied for their chemical nature, and biotic-abiotic inclusions to decipher producer linkage(s) (Kapur et al. 2024a). Further, very recently, vertebrate coprolites have also been recorded from the Miocene (Aquitanian-Langhian) interval at three locations in the Kachchh District, Gujarat State, western India (Sagar et al. 2024). Published records suggest that the Miocene (Burdigalian) Baripada beds (Odisha State, eastern India) and the late Miocene (~9 Ma) Siwalik sediments (Haritangiyal locality, Himachal Pradesh State, North India) are also known to yield vertebrate coprolites (Prasad 1968; Sharma and Patnaik 2010). Kapur et al. (2020) also provided the first evidence of Miocene (Aquitanian) microcoprolites from the Khari Nadi Formation (Kachchh region, western India. In view of the above information and considering the sporadic nature of coprolite records from India from geographically distant locations, this article is an attempt to highlight the palaeontological potential of coprolite ichnofossils, and to highlight the need for geoconservation of coprolite-yielding localities in India for future research endeavours. In the Indian scenario, coprolite ichnofossils may well be quite significant in recognizing any shift(s) in the dietary pattern(s) and ecological change(s) linked to the known climatic event(s) of the geological past as a consequence of the geodynamic evolution of the Indian subcontinent.

2 Coprolite: The Fossilized Pooh

The term "Coprolite" was coined by Buckland (1829) derived from the Greek *Kopros* meaning "dung" plus *lithos* meaning "stone" so as to describe fossilized (permineralized) faecal matter that he recovered in association with fossil skeletal remains of ichthyosaur (an extinct marine reptile) from Yorkshire, United Kingdom. Over the years, the term "coprolite" has gained wider general acceptance within the scientific community to describe fossilized (permineralized) faecal matter or pooh of animals that lived in the geological past. Coprolite are considered as trace or ichnofossils as they depict past animal activity but do not form part of the animal body (Fig. 1) just as animal footprints and borrows within the geological record.

Fig. 1 Sketch of a hypothesized terrestrial-freshwater ecosystem sustaining vegetation (palms, grasses), and fauna (titanosaurid dinosaurs, turtles, crocodiles, and fish) prevalent ~66 million years ago in central India. Note the various types of faecal matter produced and diet consumed by different animals. Sketch based on our current knowledge on coprolite data and associated biotic remains from the Deccan-volcano sedimentary sequences of central India. Original sketch by Author: Vivesh V. Kapur

Fig. 2 Late Cretaceous (Maastrichtian) striated reptilian coprolites from the infratrappean (Lameta Formation) deposits, Pisdura locality, Chandrapur District, Maharashtra State, central India. Note: Scale bar equals 1 cm for all

The parameters (e.g., size, shape, colour, chemical nature and others) of coprolites form the basis for classifying coprolites along with a general conjecture that specific groups of animals produce specifically shaped and sized excrement. In general, coprolites range in size (i.e., in longer dimensions) from a few millimetres to a few centimetres. Commonly known shapes for coprolites are not limited to elliptical, cylindrical, circular, and spiral morphotypes (Figs. 2 and 3). Various shapes depicted by these ichnofossils are a reflection of the intestinal anatomy of the producer. The classification of coprolites is still an unfinished task for palaeontologists owing to numerous factors that include (a) autonomous and diverse observations on coprolites without pursuing an apt classification system, (b) imprudent linkages to producer taxa, and (c) depiction of numerous morphological variations (Kapur et al. 2020). However, many recent attempts are quite noteworthy additions with regards to classifying coprolites (Hunt and Lucas 2012a, b; Barrios-de-Pedro et al. 2018).

Quick burial and anaerobic (absence of free oxygen) conditions are instrumental in the process of fossilization of the faecal matter, a case similar to preserve body fossils. Further, the chemical nature of the produced faecal matter governs its preservation potential. Palaeontological data on coprolites suggest that the faeces produced by the carnivores have greater preservation potential as opposed to the ones

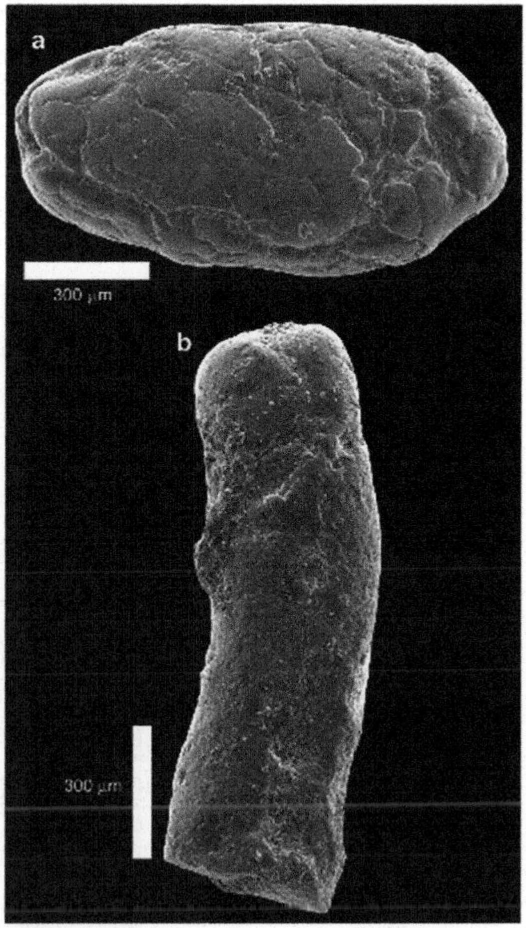

Fig. 3 (**a**) Elliptical- and (**b**) Cylindrical-shaped microcoprolites from the early Miocene (Aquitanian) Khari Nadi Formation, Kutch Basin, western India. After Kapur et al. 2020 (with permission; license no. 5912891235001)

linked to the herbivores in the geological record. This is owing to the significant role calcium and phosphate (derived from a carnivorous diet) play in pre-mineralization and subsequent lesser degree of diagenetic alteration of the faecal matter produced by carnivores (Hollocher and Hollocher 2012 and references therein; also refer to Kapur et al. 2024a).

3 Looking Within Coprolites: Dietary Clues, Internal Texture and Others

To know the diet of an animal, palaeontologists aim to look within the coprolites. For obvious reasons, coprolites of herbivorous animals preserve partly or undigested plant remains (e.g., twigs, leaf cuticles, phytoliths, spores, pollens) while that of carnivores preserve partly or undigested animal remains (e.g., bone fragments, soft tissues) (Fig. 4). In general, apart from preserving the biotic remains, the coprolites internal texture demonstrates the presence of fine-grained calcium phosphate (apatite) matrix along with the presence of thin-walled spheres (linked to sulphur producing bacteria), nano- and micron-sized porous structures formed due to escaping gases because of bacterial degradation (Kapur et al. 2020, 2024a and references therein; Fig. 4). It is important to note that not all biotic elements preserved within

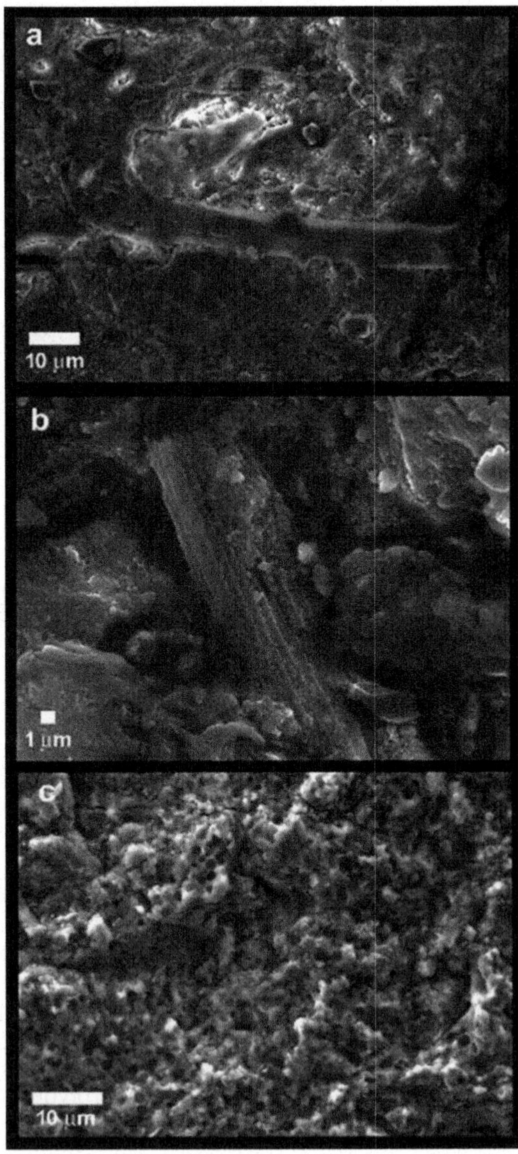

Fig. 4 Scanning electron microphotographs of a coprolite showing (**a**–**b**) partly digested food remains embedded in a calcium phosphate matrix, (**c**) micron-sized thin-walled spheres linked to sulphur producing bacteria. After Kapur et al. 2024a (with permission; license no. 5912900064119)

the coprolites reflect the main diet of the animal as it is quite likely that some elements may have been consumed accidently by the animal (see Sagar et al. 2024 and references therein). Nevertheless, direct fossil evidence within the coprolites assist to understand the past prey-predator relationship(s) and in developing the pre-historic food webs (Rakshit and Ray 2022; Kapur et al. 2024a; Sagar et al. 2024 and references therein).

4 Coprolite Research in India: Historical Background

Published records document that the coprolite research in India is more than 160 years old (Kapur et al. 2020 and references therein). The first coprolites from India were collected in 1857 from the Late Triassic (~210-million-year-old) sedimentary rocks belonging to the Maleri Formation, southern India and were documented by Oldham in 1859. Matley (1939a, b) was the first to classify (based on their size and shape) and link the Indian coprolites to their producer taxa. As presently understood, the Late Triassic coprolites recovered from the Maleri Formation have been either linked to lung fish genus *Ceratodus* or to the rhynchosaurian reptile *Paradapedon huxleyi.* Further, the Late Cretaceous (~66-million-year-old) coprolites recovered from the Deccan volcano-sedimentary successions of central India have been linked to dinosaurs, turtles and crocodiles (Kapur et al. 2024a and references therein). Coprolites attributed to crocodilians are also known from the Middle Eocene (Lutetian) Harudi Formation of Kutch Basin, western India and the Miocene (Burdigalian) Baripada beds of Orissa, eastern India (Sahni and Mishra 1975; Sharma and Patnaik 2010). Nevertheless, the ~66-million-year-old coprolites (Type-A morphotype category recovered from the Deccan volcano-sedimentary sequences of central India) linked to titanosaurid dinosaurs have received tremendous attention in the last two decades. Based on the common presence of plant tissues within Type-A coprolites, researchers argue that the Indian titanosaurid dinosaurs dominantly practised herbivory (also see Khosla et al. 2015). Apart from plant tissues, coprolites linked to titanosaurid dinosaurs are known to preserve pollen, spores, fungi, insects, diatoms, grass phytoliths, sponges and ostracods (Prasad et al. 2005; Khosla et al. 2015). Kapur et al. (2020) provided the first Indian evidence on micron-sized coprolites from the early Miocene (Aquitanian) Khari Nadi Formation, Kutch Basin, western India and attributed to ichthyophagous fishes. Recently, a diverse assemblage of microcoprolites have been recorded from the Upper Triassic Tiki Formation, central India (Kapur et al. 2024b). In separate investigations, large (cm-sized) vertebrate coprolites (linked to reptiles) have also been recorded from the Maastrichtian (~66 Ma) intertrappean deposits at Lotkheri locality, central India and from the Miocene (Aquitanian-Langhian) of the Kachchh region, western India (Sagar et al. 2024; Kapur et al. 2024a). These new discoveries highlight the palaeontological potential of these regions and hint at the future scope of research on coprolite ichnofossils in India.

5 India's Northward Journey and Changing Ecosystems

Quite crucial to India's Mesozoic-Cenozoic geological history is its northward journey (in possible isolation) after complete breakup from other Gondwanaland continents (at ~90 million years ago) to final docking with Eurasia (at ~34 million years ago), consequently forming the mighty Himalayan Mountain chain. During the course of this prolonged journey (spanning approximately 55 million years) Indian subcontinent witnessed many geological events of global significance at various time intervals. Indian subcontinent witnessed biotic changes in terms of extinctions (e.g., extinction of non-avian dinosaurs), origin and subsequently evolution of a few modern mammalian groups (e.g., perissodactyls), and intermittent introduction of large mammals within the subcontinent (e.g., gomphotheres) (Kapur et al. 2021; Kapur 2024 and references therein). These intermittent but drastic changes within the ecosystem are quite likely to have affected the dietary habits of the ancient fauna. However, to date, we have scarcely utilized coprolites associated with the Mesozoic-Cenozoic animals to understand the past ecosystems (refer to Rakshit and Ray 2022; Kapur et al. 2024a). Further, we have no or very scant knowledge in terms of any change in the diet pattern(s) of subcontinent's ancient fauna based on the coprolites due to the changes within the ecosystem at various time intervals as a consequence of severe changes in the subcontinent's climate. Therefore, coprolites can be quite instrumental in providing clues on the diet and/or changes in the dietary habit(s) linked to changes in the ancient ecosystems.

6 Gaps in Knowledge and the Way Forward

Although, research on Indian coprolites spans more than a century and a half, much emphasis has been on the ichnofossils (generally linked to dinosaurs) recovered from the ~66-million-year-old lacustrine deposits of central India. As already mentioned, we have limited knowledge of the dietary habits or preferences of diverse Mesozoic-Cenozoic faunal assemblage (particularly terrestrial vertebrates) of India. Thus, an interdisciplinary approach (incorporating palaeontological, statistical, and geochemical techniques) would be quite useful in providing insights into the changes in prey-predator interactions and dietary habits in the perspective of effects of climate change on the subcontinent's biota (refer to Rakshit and Ray 2022; Kapur et al. 2024a). The statistical and chemical characterization of Indian coprolites is incomplete, and any future studies should emphasize on utilizing various statistical tools and chemical analysis (e.g., electron dispersive spectroscopy, X-ray fluorescence, X-ray diffraction, isotopic analysis, gas chromatography-mass spectrometry) in coprolite research. Further, incorporation of non-destructive techniques such as 3D Micro-Computed Tomography (Fig. 5) should be emphasized as it provides enhanced details on the preserved biotic matter within coprolites (refer to Kapur et al., 2024a) and can help comprehend various dietary and differential digestive

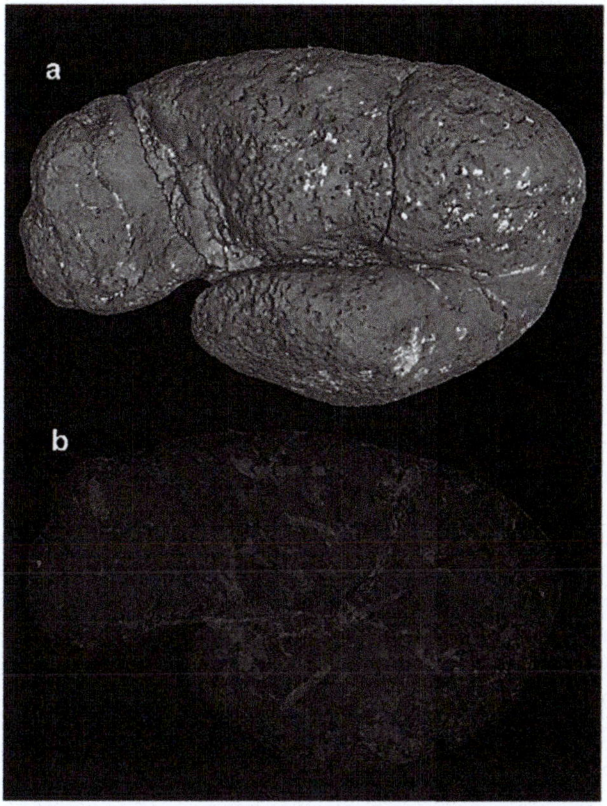

Fig. 5 Non-destructive micro-Computed Tomography utilized on Irregularly folded Morphotype 'E' reptilian coprolite specimen VVK/BNP/GEO10 (BSIP 42266) (Kapur et al. 2024a) showing (**a**) external morphology and (**b**) partly digested food remains within the ichnofossil

aspects of the producers. Micro-Computed Tomography can also be quite instrumental in data archiving, data sharing, increased social reach and appeal of these ichnofossils. This is also to bring to the reader's attention that within certain geological intervals (e.g., Oligocene), coprolite data is completely lacking from India and thus, the palaeontological efforts in the field are also warranted in addition to the laboratory research. Very recently, Kapur et al. (2024a, b) have highlighted the presence of dentalites or bite traces apart from burrows on the external surface of the Upper Triassic (Carnian-Norian) coprolites recovered from the Tiki Formation and the late Cretaceous (Maastrichtian) coprolites recovered from an intertrappean deposit. In a taphonomical context, the features preserved on the external surface of the coprolites (e.g., burrow structures, bite traces) taken together with the data on the abiotic and biotic coprolite inclusions help provide crucial inferences in terms of biotic interactions w.r.t. reconstructing past ecosystems. The above-mentioned facets also hint at the future scope of palaeontological research on coprolite ichnofossils in India.

7 Geoconservation Aspects

In the last two decades, palaeontologists conducting research in India have witnessed loss and/or destruction of numerous fossil-yielding sites owing to the lack of adequate laws, policies, lack of public awareness and related problems (Ahluwalia 2006; Kumar 2018; Banerjee 2021; Arora et al. 2023). Realizing depletion and/or loss of fossil sites, many researchers have raised concerns to safeguard the palaeontological wealth of the country through establishment of state-of-the art Museums (e.g., The Indian Museum of Earth—T.I.M.E.) that may assist in developing a better understanding in the society, generate palaeontological awareness in students and the early career researchers (Sharma 2002; Thewissen and Hughes 2010; Prasad 2012; Arora et al. 2023). Fortunately, in recent years, the United Nation protection programmes under the UNESCO's Global Geopark initiatives has gained popularity in India in view of the existence of numerous geologically significant areas/sites in India that have been identified as potential regions for geoheritage conservation (Chauhan et al. 2021; Shitole et al. 2023). Some of the potential Global Geopark regions of India such as: Bagh area (Madhya Pradesh State, central India) and Kachchh area (Gujarat State, western India) are known to yield coprolites and associated vertebrates. Particularly, the Miocene (Burdigalian-Langhian) coprolite-yielding Palasava locality has been recently proposed as a potential 'Geosite' for geoconservation and geotourism (Kapur and Chauhan 2024). Research on coprolites should be encouraged and developed in India owing to (a) the rarity of coprolite data, (b) gaps in the fossil record of these ichnofossils, (c) sporadic occurrence in geographically distant locations, and (d) the palaeontological potential of these ichnofossils. Additionally, existing and newly discovered fossiliferous sites and/or regions should be sustainably protected through geoconservation policies following an inclusive approach in accordance with the UNESCO Global Geopark concept with the local communities, state and central agencies (e.g., Forest department, Tourism department, Geoscience research and academic institutions) as the stakeholders.

8 Conclusions

Coprolite ichnofossils of pre-historic vertebrates provide crucial information on the dietary habit(s), change(s) in the dietary patterns, help to reconstruct the surrounding habitat(s), and have the potential to reveal crucial information on the pre-historic animal-environment interactions. The published coprolite records from India reveal that these ichnofossils are a rarity, gaps exist in the fossil record of these ichnofossils from various time-slices (e.g., Oligocene) and these ichnofossils occur sporadically in geographically distant locations of India. Thus, considering the palaeontological potential of coprolite ichnofossils urgent and necessary geoconservation of coprolite-yielding localities in India is warranted. Ongoing geoconservation efforts in India can be instrumental to sustainably protect the existing and the newly discovered fossiliferous (coprolite-yielding) sites and/or regions. Further, the standard protection measures should be implemented following an inclusive approach with local communities, state and central government agencies (e.g., Forest Department, Tourism Department, Geoscience research and academic institutions) as the stakeholders.

Acknowledgements The author would like to thank Dr. Vandana Prasad (Ex-Director, BSIP) for encouragement and necessary permission(s). The author is grateful to Professor MG Thakkar (Director, BSIP) for constant support and encouragement. The author thanks Dr. S C Tripathi (SES, Lucknow) for the invitation to contribute. The author also sincerely acknowledges Professor Ashok Sahni (Emeritus Scientist, Panjab University, India) for constant encouragement and support to research Indian coprolite material and for the valuable discussion(s) on the subject matter. Funding support in the form of BSIPs In-house Project 3 (2021-2025) and SERB (CRG) Grant no. CRG/2019/002204 is thankfully acknowledged. The author is grateful to both the reviewers and the editor(s) for constructive commentaries on the manuscript.

References

Ahluwalia AD (2006) Indian geoheritage, geodiversity: geosites and geoparks. Curr Sci 91:1307

Arora P, Trivedi PM, Bhatia H, Agnihotri P, Kapur VV (2023) A survey of the anticipated role of the Indian Museum of Earth (TIME) to Foster public awareness towards the preservation of Palaeontological relics. Geoheritage 15(109). https://doi.org/10.1007/s12371-023-00877-y

Banerjee DM (2021) Fate of Indian geoheritage and geopark bill. Curr Sci 121(2):183–184

Barrios-de Pedro S, Poyato-Ariza FJ, Moratalla JJ, Buscalioni ÁD (2018) Exceptional coprolite association from the early cretaceous continental Lagerstätte of Las Hoyas, Cuenca, Spain. PLoS One 13(5):e0196982. https://doi.org/10.1371/journal.pone.0196982

Buckland W (1829) On the discovery of a new species of Pterodcatyle; and also of the faeces of the ichthyosaurus; and of a black substance resembling sepia, or India ink, in the Lias at Lyme Regis. Proc Geol Soc Lond 1:96–98

Chauhan G, Biswas SK, Thakkar MG, Page KN (2021) The unique geoheritage of the Kachchh (Kutch) Basin, Western India, and its conservation. Geoheritage 13(1):1–34

Hollocher K, Hollocher TC (2012) Early processes in the fossilization of terrestrial feces to coprolites, and microstructure preservation. In: A.P. Hunt, J. Milàn, S.G Lucas, and J.A. Spielmann (eds.), Vertebrate coprolites. New Mexico Museum of Natural History and Science, Bulletin 57:79–92

Hunt AP, Lucas SG (2012a) Classification of vertebrate coprolites and related trace fossils. In: A.P. Hunt, J. Milàn, S.G Lucas, and J.A. Spielmann (eds.) Vertebrate coprolites. New Mexico Museum of Natural History and Science, Bulletin 57:137–146

Hunt AP, Lucas SG (2012b) Descriptive terminology of coprolites and recent feces. In: A.P. Hunt, J. Milàn, S.G Lucas, and J.A. Spielmann (eds.), Vertebrate coprolites. New Mexico Museum of Natural History and Science, Bulletin 57:153–170

Kapur VV, Chauhan G (2024) Miocene (~ 14 ma) vertebrate-yielding site of Palasava, Kachchh, Gujarat State, Western India: geoheritage-geotourism perspectives and geosite potential. Geoheritage 16:12. https://doi.org/10.1007/s12371-024-00915-3

Kapur VV, Bajpai S, Sarvanan N, Das DP (2006) Vertebrate fauna from Deccan Intertrappean beds of Bhanpura, Mandsaur District, Madhya Pradesh. Gondwana Geol Magaz 21:43–46

Kapur VV, Kumar K, Morthekai P, Chadha A (2020) Palaeodiet of Miocene producer(s) and depositional environment(s): inferences from the first evidence on microcoprolites from India.

Acta Geol Sinica-English Edition. Acta Geologica Sinica - English Edition 94(5):1574–1590. https://doi.org/10.1111/1755-6724.14293
Kapur VV, Sagar R, Kumar K, Chaddha AS, Lourembam RS, Mishra A, Sharma A (2024a) Palaeobiological and geochemical aspects of reptilian coprolites from a Maastrichtian Deccan volcano-sedimentary intertrappean deposit of central India. Ichnos 31: 1–24. https://doi.org/10.1080/10420940.2024.2415153.
Kapur VV, Sagar R, Singh K, Prasad GVR (2024b) A new microcoprolite assemblage from the Upper Triassic Tiki Formation of India: ichnotaxonomy, and producer association(s). Historical Biology 1–15. https://doi.org/10.1080/08912963.2024.2402263.
Khosla A, Chin K, Alimohammadin H, Dutta D (2015) Ostracods, plant tissues, and other inclusions in coprolites from the late cretaceous Lameta formation at Pisdura, India: Taphonomical and palaeoecological implications. Palaeogeogr Palaeoclimatol Palaeoecol 418:90–100
Kumar S (2018) Edge of extinction. Science 360(6384):22–25. https://doi.org/10.1126/science.360.6384.22
Lindlahr VH (1942) You are what you eat, 5th edn. National Nutrition Society, p 128
Matley CA (1939a) On some coprolites from the Maleri beds of India. Rec Geol Surv India 74:530–534
Matley CA (1939b) The coprolites of Pisdura, central province. Rec Geol Surv India 74:535–547
Oldham T (1859) On some fossil fish-teeth of the genus ceratodus, from Maleri, south of Nagpur. Memoirs Geol Surv India 1:295–309
Prasad KN (1968) The vertebrate fauna from the Siwalik beds of Haritalyangar, Himachal Pradesh, India. Palaeontol Indica XXXIX:1–78
Prasad GVR (2012) Need for a national prehistoric museum. Curr Sci 102(7):955
Prasad V, Stromberg CAE, Alimohammadian H, Sahni A (2005) Dinosaur coprolites and the early evolution of grasses and grazers. Science 310:1177–1180
Qvarnström M, Vikberg Wernström J, Wawrzyniak Z, Barbacka M, Pacyna G, Górecki A, Ziaja J, Jarzynka A, Owocki K, Sulej T, Marynowski L, Pieńkowski G, Ahlberg PE, Niedźwiedzki G (2024) Digestive contents and food webs record the advent of dinosaur supremacy. Nature 636: 397–403. https://doi.org/10.1038/s41586-024-08265-4.
Rakshit N, Ray S (2022) Bone-bearing coprolites from the Upper Triassic of India: ichnotaxonomy, probable producers and predator-prey relationships. Papers in Palaeontology 8: e1418. https://doi.org/10.1002/spp2.1418.
Sagar R, Kapur VV, Kumar K, Morthekai P, Sharma A, Shukla S, Ghosh AK, Chauhan G, Thakkar MG (2024) The first record on cm-sized vertebrate coprolites from the early-middle Miocene (Aquitanian-Langhian) Khari Nadi and Chassra formations, Kutch Basin, western India: palaeobiological significance. Pre-print SSRN electronic journal. https://doi.org/10.2139/ssrn.4269941
Sahni A, Mishra VP (1975) Lower tertiary vertebrates from western India. Monogr Palaeontol Soc India 3:1–48
Sharma M (2002) Palaeontology in India at crossroads. Curr Sci 82(8):913–917
Sharma MK, Patnaik R (2010) Coprolites from the lower Miocene Baripada beds of Orissa. Curr Sci 99(6):804–808
Shitole AD, Gupta V, Gidwani L, Verma V, Rajora S, Anand S, Tripathi SC (2023) Bagh dinosaur National Park Region, Dhar District, Madhya Pradesh: a potential UNESCO global Geopark. Geoheritage 15:116. https://doi.org/10.1007/s12371-023-00881-2
Thewissen JGM, Hughes N (2010) Save India's heritage. Curr Sci 99(3):267

Chitrakoot Region, Madhya Pradesh and Uttar Pradesh: A Potential UNESCO Global Geopark

Satish C. Tripathi, Anil Kumar, and Pranay Kant Biswas

Abstract India's geological history has created exceptional geodiversity, featuring stable cratons, dynamic mountains, vast volcanic provinces, and diverse sedimentary basins that shape its unique landscape. Despite rich geological diversity and the conservation efforts directed towards geoheritage sites, the establishment of national or UNESCO Global Geoparks remains a pending endeavor. This paper aims to assess the potential of the Chitrakoot region in Madhya Pradesh and Uttar Pradesh to develop into a geopark, serving as a model for similar regions in India. The region's geological significance is notable, featuring the unique contact of Vindhyan sediments with its basement Bundelkhand granites exposed only in Chitrakoot in Central India. Majhagawan, the only working diamond mine in India, wildlife centuries, exotic waterfalls, caves, rock paintings, and historic sites make it very close to a geopark. The proposed Chitrakoot Geopark satisfies the criteria for global geopark status, necessitating actions for UNESCO recognition. This endeavor not only promotes community-led geoheritage conservation but also fosters sustainable development through local handicrafts and traditions.

Keywords Vindhyan · Bundelkhand · Geodiversity · UNESCO Global Geoparks · Stromatolites

S. C. Tripathi (✉)
The Society of Earth Scientists, Lucknow, Uttar Pradesh, India

A. Kumar
Department of Geography, D.S.N (P.G.) College, Unnao, Uttar Pradesh, India

P. K. Biswas
Department of Geography, C.M.P. College, University of Allahabad, Prayagraj, Uttar Pradesh, India

S. C. Tripathi et al. (eds.), *Geoconservation and Geotourism Potential of India*, Society of Earth Scientists Series, https://doi.org/10.1007/978-3-031-81017-6_12

1 Introduction

India is known for its rich geological diversity, straddling from the Archean Era to the Quaternary Period. This geological diversity has played a significant role in shaping the country's varied landscapes and has contributed to its geomorphological and cultural diversity. Geoheritage encompasses global, national, provincial, and local features of geology at all scales that are inherently important sites or culturally important sites offering information or insights into the evolution of the Earth, into the history of science, or that can be used for research, teaching, or reference (Brocx and Semeniuk 2007). Geoheritage sites can be classified as geological, geomorphological, and geoarchaeological heritage sites as per their origin and characteristics. The geological, geomorphological, and cultural diversity of India contributes to its uniqueness and provides a wealth of opportunities for scientific study, tourism, and sustainable development.

Although concepts like geoparks, geodiversity, geoheritage, geoconservation, and geotourism are gaining global acceptance, these are relatively new in India. Geological Survey of India declared 32 geoheritage sites as National Geological Monuments (NGMs) (Anantharamu et al. 2001 and later) at different times, but except for a few fossil sites, others are in dilapidated condition. Similarly, there are 213 UNESCO Global Geoparks in 48 countries as of 2024. But India has none, whereas there are 42 World Heritage Sites. Several geoheritage regions in India can be converted and promoted as National/ UNESCO Global Geoparks (Shitole et al. 2023). The untapped geodiversity and geoheritage in India need effective presentation both nationally and internationally. Several regions in India, meeting UNESCO Global Geopark norms, have the potential for recognition but lack a structured approach for documentation, presentation, and promotion to attract national and international attention. A team of knowledgeable and active geoscientists drafted an elaborate document entitled 'Conservation of Geoheritage Sites and Development of Geoparks Act-2020' (Banerjee 2021). However, such legislation has not seen the light of day.

This paper aims to document and evaluate the potentiality of establishing a Geopark in the Chitrakoot region which falls in the states of Madhya Pradesh and Uttar Pradesh. The case study will prompt other similar geoheritage regions to be considered and evaluated for designation as a Geopark.

2 Chitrakoot Region: A Potential Geopark

The proposed Chitrakoot Geopark is situated in the Bundelkhand region of central India and is administratively spread over the states of Madhya Pradesh and Uttar Pradesh (Fig. 1). The transitional location of the geopark determines its subtropical monsoon rhythm. The temperature varies from 2.5 °C (January) to 50.8 °C (May) in this region (Verma 2020). This variation is a result of its rocky surface and scanty

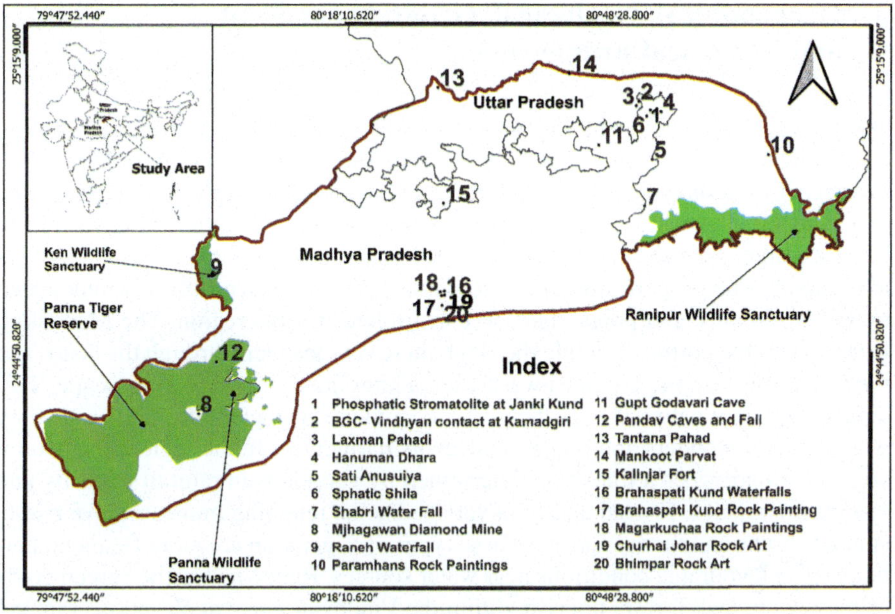

Fig. 1 Location and boundary of the proposed Chitrakoot Geopark

vegetation. Casual and irregular changes in temperature are followed by freaks of wind and weather. Here, rainfall is caused by both the Bay of Bengal current and the Arabian Sea current. Wherever the two currents interact, heavy rain is caused. The Vindhyan and Panna ranges make obstructions in the passage of the southeast monsoon current, which causes heavy rainfall in Panna (the southwestern part of the geopark). The average rainfall for the geopark region is 116 cm. Three wildlife sanctuaries (and two tiger reserves) are situated in this region: Ranipur Wildlife Sanctuary (Ranipur Tiger Reserve), Ken Wildlife Sanctuary, Panna Tiger Reserve, and Panna Wildlife Sanctuary. The region contains excellently preserved geomorphological, geological, geoarchaeological, historical, ecological, and cultural sites attesting to the concept of a geopark.

The word "Chitrakoot" translates from 'hill of paintings'; the home to numerous rock paintings in caves and rock shelters. The Kalinjar Fort's rock-cut sculpture is a significant indicator of the geoarchaeological framework. As a living museum, the 'Chitrakoot Potential Geopark' invites us to discover and educate ourselves about the value of cultural and natural riches.

3 Geographical, Geological, Geoarchaeological and Ecological Framework

The proposed geopark, spanning an area of 4377 km^2 with a perimeter of 643 km, encompasses regions within Madhya Pradesh and Uttar Pradesh, specifically the districts of Chitrakoot and Banda in Uttar Pradesh and Chhatarpur, Satna, and Panna in Madhya Pradesh. The proposed geopark is bordered by National Highway-35 to the east, Panna Tiger Reserve to the west, the Ken River to the north, and Baghelkhand to the south. The geopark stretches across five districts and is beautifully punctuated by rivers, streams, and ponds that serve as lifelines for the region. The Mandakini River graces the northeast, while the Baghain River meanders through the heart, and the Ken River flows in the northwest, each adding its charm to the landscape. This delicate dance of nature shapes the geopark's unique character, offering a sanctuary where natural wonders and human endeavors harmonize. In the region, the natural vegetation undergoes a transition determined by variations in rainfall intensity and soil characteristics. Eroded and impoverished soils, retaining moisture, give rise to stunted shrubs and thorny forests. These types of vegetation are more frequent, particularly in favourable conditions near water sources. Renowned as the "green gold" of the region, teak forests flourish within the Vindhyan range at elevations ranging from 400 to 500 meters above sea level. This species holds significant economic value. At lower elevations, between the teak forests and riverine tracts, mixed forests thrive. These forests encompass a diverse array of plant species, including *tendu* (*Diospyros melanoxylon*), *khair* (*Acacia catechu*), and *achar* (*Buchanania lanzan*). *Tendu* leaves are commonly used in the production of *bidis*; *khair* is valued for its catechu extract; and *achar* may denote *Buchanania lanzan*, a tree known for its edible fruits. In the semi-arid low-waste lands near Ken and its tributaries, uncontrolled grazing has led to a deplorable state, with overgrazing and vegetation loss. Karonda, Karil, Babul, Jharberi, Reonja, and Dhak endure this challenging environment alongside key species like *Zizyphus*, *Calotropis, Prosopis, Capparis,* and *Dichrostachys*, highlighting the resilience of local flora.

The proposed geopark is geologically composed of the sedimentary rocks of the Bijawer Group, Vindhyan Supergroup along with the basement Bundelkhand Granitoid Complex. Vindhyan sedimentary basin, which is an intracratonic depositional domain and comprises a thick sedimentary succession (about 4500 m) consisting of conglomerate, sandstone, shale, limestone/dolostone, and volcaniclastics and intrusive kimberlites/lamproites (Ahmad 1971).

The paleontological framework of Chitrakoot Geopark reveals a rich geological history. Fossils from the Vindhyan Supergroup provide insights into early life forms. The region's sedimentary rocks preserve an ancient ecosystem where red and blue-green algae thrived, which were responsible for changing the atmosphere for the future growth of life. Earth's oldest red algae was discovered from stromatolite samples from Janki Kund (Bengtson et al. 2017).

The ecological framework is the most important aspect of a geopark. UNESCO Global Geoparks (UGGPs) are seen as synonymous with environmental protection and development (Patzak and Eder 1998). Chitrakoot Geopark lies in a vibrant ecosystem teeming with life. Here, amidst the lush landscapes and winding rivers, two tiger reserves, Panna and Ranipur, stand as guardians of the majestic big cats, ensuring their safety and survival. These reserves are not just havens for tigers but also bustling hubs of biodiversity, where countless species find refuge and thrive. Alongside the tiger reserves, a sanctuary is dedicated to protecting the 'gharial', a rare and endangered crocodilian species. These gentle giants roam the waters; their presence is a testament to the importance of conservation efforts in safeguarding vulnerable species. And amidst the verdant forests and rolling hills lies Panna National Park, a park for a diverse array of wildlife. Here, amidst the rustling leaves and chirping birds, antelopes frolic in the grasslands while leopards prowl in the shadows. Together, these protected areas form a tapestry of life within the Chitrakoot Geopark, each playing a vital role in preserving the natural beauty and biodiversity of the region. The proposed Geopark exemplifies the need for harmony with nature, urging visitors to cherish and protect its pristine landscapes and natural wonders for future generations. The village-forest dynamics of Panna National Park within the Chitrakoot Geopark would leverage the interdependence of village-forest relations towards more sustainable outcomes for all stakeholders (Runacres 2020).

Chitrakoot, the key feature of the proposed geopark known as the 'Hills of Many Wonders', holds immense cultural and geological significance. It was revered as Lord Ram's sacred abode during his exile and boasts unique geological formations, including the exposed contact between Bundelkhand granite and Vindhyan sediments, making it a compelling destination for scholars and pilgrims alike. Janki Kund at Chitrakoot contains phosphatic stromatolite fossils, *Ramathallus lobatus*, and *Rafatazmia chitrakootensis*, dating back to 1600 million years, crucial for Earth's oxygenation. Adjacent to Chitrakoot, Majhgawan-Panna hosts India's sole diamond mine and the renowned Panna Tiger Reserve. Raneh Canyon offers a unique geological landscape with excellent rock exposure. Table 1 provides details of the tentatively identified geosites of the proposed Chitrakoot Geopark.

Some of the significant geosites are detailed below:

1. BGC—Vindhyan contact at Kamadgiri

This site is located at the border of the Chitrakoot district of Uttar Pradesh and the Satna district of Madhya Pradesh. The central India is largely composed of Bundelkhand granitoids and the overlying Vindhyan Supergroup of rocks (Fig. 2). Kamadgiri is best location in Central India where we are able to see the contact of Vindhyan rocks (Semri Group; more than 1600 My) with its basement, the Bundelkhand Granite Complex (BGC; 3500–2500 My). The contact represents about 900 million years of hiatus (the status quo in the history of Earth) in Earth's evolution. The hill bears a sacred cultural value as it was the abode of Lord Ram for over 11 years during the period of his exile. Therefore, it is geoscientifically as well as culturally a site of international significance.

Phosphatic Stromatolites of Janaki Kund

The Janki Kund, said to be the bathing place of goddess Sita, is full of fossils of red and blue-green algae known as stromatolites, one of the oldest life forms on Earth

Table 1 List of geosites (marked in Fig. 1) for the proposed 'Chitrakoot Geopark'

No.	Name	Latitiude	Longitude	Geological, Geomorphological, Geoarchaeological and Cultural Significance
1.	Phosphatic stromatolite at Janki Kund	25.15768	80.86356	Janki Kund in Chitrakoot exposes phosphatic stromatolite fossils, such as *Ramathallus lobatus* and *Rafatazmia chitrakootensis*, dating back 1600 million years, vital for Earth's oxygenation
2.	BGC-Vindhyan contact at Kamadgiri	25.17244	80.84279	Kamadgiri uniquely exhibits the contact between Bundelkhand granite with overlying Vindhyan sediments. It is the only place where this contact is exposed
3.	Laxman Pahadi	25.16516	80.8377	Lakshman hill near Kamadgiri, Chitrakoot, houses temples and offers scenic views, including a sunset spot and Laxman Kund
4.	Hanuman Dhara	25.15503	80.88322	A pilgrimage site with a temple dedicated to lord Hanuman and a perennial stream flowing over the deity
5.	Sati Anusuiya	25.07405	80.86705	Sati Anusuiya temple, the origin of Paisuni River (locally Mandakini), features an escarpment of Kaimur Sandstone
6.	Sphatic Shila	25.14721	80.85676	The religious mythological site features alternating chert and dolomite bands, shaped by changing depositional conditions and pressure-solution processes
7.	Shabri Water Fall	24.99518	80.85305	Shabri waterfall offers geological marvels and mythological significance. Enhanced infrastructure and security measures attract growing tourist numbers
8.	Majhagawan Diamond Mine	24.64287	80.03301	The primary source open-cast mine yields economically precious diamonds from lamprophyre/kimberlite, significant in the world's diamond industry
9.	Raneh Waterfall	24.8758	80.05126	Ken River carves a 5 km long canyon of colorful phases of granite, featuring perennial and seasonal waterfalls, a stunning natural spectacle
10.	Paramhans Rock Paintings	25.08185	81.08173	Geoarchaeological heritage site rock paintings of the hunter-gatherers community are found here
11.	Gupt Godavari Cave	25.09793	80.76879	A set of Vindhyan Tirohan limestone caves is geologically significant with river-like flowing water
12.	Pandav Caves and Fall	24.73068	80.06655	30-meter cascade into a heart-shaped pool, surrounded by lush forests. Serene, mystical, with ancient caves; believed Mahabharata Pandavas' refuge

(continued)

Table 1 (continued)

No.	Name	Latitiude	Longitude	Geological, Geomorphological, Geoarchaeological and Cultural Significance
13.	Tantana Pahad	25.19583	80.47472	Ringing Bell Rock, Naraini Banda, UP
14.	Mankoot Parvat	25.21945	80.71525	Archaeological site houses medieval Shivlings, offering a glimpse into ancient religious practices and cultural heritage
15.	Kalinjar Fort	24.9997	80.4852	Kalinjar Fort, Banda, Uttar Pradesh, built by Chandela Kings, boasts impressive architecture, cultural richness, and strategic importance, attracting tourists worldwide
16.	Brahaspati Kund Waterfalls	24.84392	80.48713	Brihaspati Kund, Panna, Madhya Pradesh, a natural waterfall in Bundelkhand, Madhya Pradesh
17.	Brahaspati Kund Rock Painting	24.84306	80.48194	The ceilings of the shelters are depicted with pictographs of hunting, agriculture, domestication of livestock, and wars in the Iron Age to as early as the Mesolithic age (Dehuri and Pradhan 2018)
18.	Magarkuchaa Rock Paintings	24.84917	80.47944	Magarkachha rock paintings depict faunal activities, social life, and cultural evolution from Mesolithic to historical periods
19.	Churhai Johar Rock Art	24.85056	80.48694	Rock paintings, located near Brihaspati Kund, 15 m long, and 7 m high, mainly battle scenes, horse riders, and handprints. Human figures are depicted with thin lines and stick-like patterns
20.	Bhimpar Rock Art	24.826777	80.482916	Near Brihaspati Kund, Gajna village, named after Bhimparhar hillock, features rock paintings, natural shelters, and Bhim Kund springs

Fig. 2 Contact between Bundelkhand granitoids (basement) and overlying Vindhyan sediments (Semri Group) at Kamadgiri

Fig. 3 Phosphatic stromatolites of Janki Kund, Chitrakoot exposed along Mandakini River

(Fig. 3). The Swedish Museum of Natural History in Stockholm studied fossils from Chitrakoot and found two types of multicellular colony dating to 1.6 billion years ago. One is a thread-like form (*Rafatazmia chitrakootensis*), whereas the other is lobe-shaped (*Ramathallus lobatus*) (Bengtson et al. 2017). These algae were responsible for increasing the amount of oxygen in the atmosphere by consuming carbon dioxide and releasing oxygen.

Gupt Godavari Limestone Caves

The Gupt Godavari group of caves are developed in Tirohan limestone (Fig. 4). Two caves are linked together and a third cave has been identified during the fieldwork which also appears to be linked with the other two caves. Limestone solution caves are common in India but Gupt Godawari caves are unique in the sense that they are full of running water like a river which is called Godawari. Ancient Hindu literature also describes the existence and use of these caves.

Shabri Waterfall

Shabri Waterfall is located on the outskirts of Chitrakoot (Fig. 5). The most picturesque waterfall is developed along the Paisuni River close to Markundi town in the Manikpur forest (Ranipur Wildlife Sanctuary). It is commonly visited by tourists, mainly during the rainy season. It also marks the state boundaries of Uttar Pradesh and Madhya Pradesh. The state governments have built the first 'Glass Skywalk Bridge' to attract visitors to see the natural beauty of the valley. Several mythological stories are associated with this waterfall.

Ringing Rocks of Tantana Pahad

This geoheritage site is located in Naraini Nagar Panchayat of Banda District, 35 km. south of Banda City and 53 km. west of Chitrakoot. There is a balancing rock (granite) that rings like a bell when you hit it, locally named 'Tantana Pahad'

a b

Fig. 4 (**a**) Gupt Godawari cave. (**b**) Erosion pattern of Tirohan limestone within the cave

Fig. 5 Panoramic view of Shabri waterfall

(Fig. 6). Tantana Pahad has a fascinating geological oddity similar to ringing rocks seen around the world. The lithophonic marvel is a place of attraction for visitors.

Kalinjar Fort

Kalinjar Fort stands on a flat-topped hill consisting of Kaimur sandstone, situated in Banda District, Uttar Pradesh. It was ruled by several dynasties, including the Guptas, the Vardhana dynasty, the Chandelas, the Solankis of Rewa, the Mughals, and the Marathas. It has also been mentioned in Buddhist literature, particularly in the travelogues of Buddha. The Chedi dynasty ruled here during the time of Gautama Buddha (563–480 BC). The third to fifth-century Kalbhairav rock cut (Fig. 7) is an architectural marvel made up of sandstone and granite. Kalinjar serves as a strategic fortress whose walls are adorned with rare stone sculptures as well as intricate carvings of Hindu gods and deities, as well as legendary beings' exquisiteness, portraying its real significance as a historical fort-cum-temple reflecting the era over which it was built. The strategic location of Kalinjar Fort at the end of the Vindhyan range, overlooking the Bundelkhand plains, has always made it a prominent place in history.

Fig. 6 Ringing balancing rock of Tantana pahar, Banda

Fig. 7 Kalbhairav rock cut of Kalinjar Fort, Banda

Fig. 8 Raneh Canyon along Ken River near Khajuraho, Madhya Pradesh

Raneh Canyon
Raneh Canyon and Falls, situated in Khajuraho, Madhya Pradesh, showcases a 5-km-long and 30-meter-deep canyon along the Ken River, exposing a variety of Bundelkhand granite and making the canyon colourful (Fig. 8). It also gives a good opportunity to study the various phases of granite. During the monsoon season, seasonal waterfalls further boost its beauty. The proximity of the Ken Gharial Sanctuary and Pandav Falls further enriches the nature's experience.

Majhagawan Diamond Mine
The open-cast diamond mine of Majhagawan, owned by National Mineral Development Corporation, started production in 1971–72 (Fig. 9). It is the only working diamond mine in India. It is located about 15 km from the Panna. The deposit is in a kimberlites/lamproites pipe spread over a 6.5 ha area. The Panna Tiger Reserve is another added attraction within the proposed geopark region.

4 Crafts Associated with Geopark

The Wooden Craft of Chitrakoot Chitrakoot district has a distinct place in the country in woodcraft. A good mixture of the traditional way and modernity was done, and the craftsmen of the district added new horizons to this art (Fig. 10). Being a forest area, wood (raw material) is abundant in Chitrakoot. A large number of craftsmen in this district are engaged in making wooden toys. The toys produced here are sent to different districts of the state for sale as well as for fairs and exhibitions. Wood crafts are promoted under the 'One District, One Product' scheme of the government of Uttar Pradesh.

Fig. 9 Majhgawan diamond mine, Panna, Madhya Pradesh

Fig. 10 wooden crafts of Chitrakoot developed under One District, One Product scheme of UP

The Shazar Stone Craft of Banda Sazar stones of Banda are agates with dendrites occurring in the Ken River bed (Fig. 11). These semi-precious stones were discovered about 400 years ago and are exported mainly to Arab countries. Other than agate, decorative and polished pebbles, and stones are another commercial product of Banda. Though the business currently is not in a good state but creation of Geopark will boast a rich historical tradition.

Metal Craft of Chhatarpur Metal crafting in Chhatarpur, Madhya Pradesh, is another ancient art form carved on ordinary brass vessels (Fig. 12). The delicately

Fig. 11 Shajar stone craft of Banda

Fig. 12 Metal crafting in Chhatarpur, Madhya Pradesh

designed figures of gods, goddesses, animals, and abstract patterns on the surface of the metal attract the buyers. Now the art is further improving by treating the brass with copper, silver, and nitric acid to achieve an antique finish, enhancing the depth and beauty of the artwork.

5 Geopark Assessment Criterion

The Chitrakoot region holds great potential for development into a recognized 'Chitrakoot Geopark' that could earn a place on UNESCO's list. This assessment explores the region's alignment with UNESCO's standards. Meeting all the requirements for a prospective UNESCO Geopark, Chitrakoot stands out as a prime contender. However, proper documentation and development on the concept of a geopark are yet to be done. We provide here a basic assessment and potentiality of the Chitrakoot region; however; further detailed blueprint and action plans are to be prepared.

Geological Heritage of International Value The Chitrakoot Geopark proposal is notable for its internationally valuable geological heritage. Geologically, Chitrakoot is the only location in central India where we can study exposed contact of Bundelkhand granitoids with the overlying Vindhyan sedimentary basin, a hiatus of a 900 million-year. The ~1.6 Ga Tirohan dolomite of the Lower Vindhyan in central India contains phosphatized stromatolitic microbialites. The fossil red algae have been considered the oldest on Planet Earth (Bengtson et al. 2017).

The phosphatic stromatolites exposed in Janki Kund are among the oldest known life forms on Earth. Paleontologists identified globular structures interpreted as gas bubbles in cyanobacterial biomats from Janaki Kund stromatolites (Sallstedt et al. 2018). The international scientific community has honored the site's connection to Lord Ram by naming these fossils *Ramathallus lobatus* and *Rafatazmia chitrakootensis*. These ancient algae played a crucial role in transforming our planet's atmosphere by absorbing carbon dioxide and releasing oxygen, which helped life on Earth to flourish.

The Majhgawan diamond mine of the Panna district of Madhya Pradesh is the only producer mine of diamonds in India. The kimberlite/lamproite pipe intruding Vindhyan sediments yielded diamonds which were commercially exploited. It is the geologically most significant site of the proposed Geopark.

Management The establishment of the Chitrakoot Geopark, encompassing five districts across Uttar Pradesh and Madhya Pradesh, requires a cohesive management framework. This framework should integrate the governance of the whole region into one geopark administrative body. There are two tiger reserves, a gharial reserve, and three wildlife sanctuaries. The tiger reserves are under the purview of the National Tiger Conservation Authority, and the gharial sanctuary is protected under the Wildlife Protection Act of 1972 by the State Forest departments of both states. The proposed geopark, which includes major urban centers such as Karvi, Chitrakoot Nagar Parishad, and Panna, necessitates the creation of a singular geopark governing body. This body will ensure unified management, aligning conservation efforts with sustainable development and community involvement within the geopark's jurisdiction.

Legal Protection The proposed Chitrakoot Geopark encompasses a mosaic of geosites, many of which are nestled within areas of environmental and cultural protection. Notable among these are the Raneh Canyon within the Ken Gharial Sanctuary and the Pandav Falls situated in the Panna Wildlife Sanctuary. Additionally, sites of historical significance such as the Kalinjar Fort and Mankoot Parvat, protected by the Archaeological Survey of India, contribute to the region's rich tapestry of heritage. The stewardship of most geosites falls under the jurisdiction of the State Forest Department, while others are intertwined with the spiritual fabric of the area, being managed by local religious communities-examples include Kamadgiri,

Lakshman Pahadi, Hanuman Dhara, and Sphatic Shila. However, certain sites like the rock paintings at Paramhans Ashram and Tantana Pahad are yet to receive adequate protection, highlighting an area of concern and opportunity for conservation efforts. This potential can be actualized through concerted efforts by the Government of India and the state governments of Uttar Pradesh and Madhya Pradesh, emphasizing the need for a collaborative approach to celebrate and preserve this repository of natural and cultural wealth.

Visibility The Chitrakoot region exemplifies the intricate interplay between geography, geology, and religious beliefs. This confluence has laid the foundation for numerous narratives that are deeply ingrained in the collective consciousness of India's populace. These stories, borne out of the region's unique assimilation of natural and spiritual elements, are reflected in the everyday lives and cultural practices of its people. The Chitrakoot region, steeped in the lore of Vanvasi Ram and Pandavas and graced by the natural sanctuaries of Panna Tiger Reserve and Ken Gharial Reserve, is also home to the historic Kalinjar Fort. This area is not only a confluence of cultural and natural heritage but also a site of profound scientific interest. The seminal work by Sallstedt et al. (2018) unveiled the presence of oxygenic phototrophy in ancient phosphatic stromatolites from the region, underscoring the region's significance in the narrative of early life on Earth. These discoveries collectively highlight the Chitrakoot region's multifaceted importance as a cradle of biodiversity, a bastion of history, and a crucible of geobiological research. It is this synthesis that not only enriches the region's identity but also contributes to its potential as a UNESCO Global Geopark, where such interconnectedness can be studied, preserved, and celebrated.

Education The Chitrakoot region stands as a testament to the educational richness embedded in its geological and cultural landscapes. It serves as a natural classroom, illustrating the first evidence of life formation on Earth. The ancient stromatolites at Janki Kund exemplify early life forms that contributed to atmospheric oxygenation, while Kamadgiri (the abode of Lord Ram) is the best location for the only exposed contact of lower Vindhyan sediments and basement Bundelkhand granite in Chitrakoot. The contract represents about 900 My long hiatus (the status quo in the history of Earth) in Earth's evolution. The region's dry deciduous forests, combined with bright native art, handicrafts, and a distinct tribal lifestyle, add to the area's rich tapestry. This combination of natural beauty and cultural diversity makes the Chitrakoot region an excellent candidate for designation as a UNESCO Global Geopark.

References

Ahmad F (1971) Geology of the Vindhyan system in the eastern part of the Son Valley, Mirzapur District, U.P. Rec Geol Sur India 96. 41 p

Anantharamu TR, Bellur D, Bhasker AA (2001) National geological monuments. Geol Surv India Spec Publ 61. 98 p

Banerjee DM (2021) Fate of Indian Geoheritage and Geopark Bill. Curr Sci 121(2):183–184. https://doi.org/10.14712/23361980.2023.10

Bengtson S, Sallstedt T, Belivanova V, Whitehouse M (2017) Three-dimensional preservation of cellular and subcellular structures suggests 1.6 billion-year-old crown-group red algae. PLoS Biol 15(3):e2000735

Brocx M, Semeniuk V (2007) Geoheritage and geoconservation-history, definition, scope and scale. J R Soc Western Austr 90:53–87

Dehuri R, Pradhan A (2018) Re-visiting Brihaspati kund and discovery of rock art in its adjoining area in district-Panna, Madhya Pradesh. ICON J Archaeol Cult 5:125–132

Patzak M, Eder W (1998) UNESCO Geopark: a new programme, a new UNESCO label. Geol Balc 28:33–36

Runacres A (2020) Engaging conservation: village-Forest relations around Panna Tiger Reserve in Central India. Ph.D. thesis. Department of Anthropology, University College London

Sallstedt T, Bengtson S, Broman C, Crill PM, Canfield DE (2018) Evidence of oxygenic phototrophy in ancient phosphatic stromatolites from the Paleoproterozoic Vindhyan and Aravalli Supergroups, India. Geobiology 16(2):139–159. https://doi.org/10.1111/gbi.12274

Shitole AD, Gupta V, Gidwani L, Verma V, Rajora S, Anand S, Tripathi SC (2023) Bagh dinosaur National Park Region, Dhar District, Madhya Pradesh: a potential UNESCO global Geopark. Geoheritage 15(4):116. https://doi.org/10.1007/s12371-023-00881-2

Verma BL (2020) Geography of Bundelkhand. Raj Publication. 294p

Geotourism Potential of Quaternary Colluvio-Aeolian Deposits and Unique Parabolic Dunes Landscape in Central Thar Desert of India: Insights into Geoheritage Sites Depicting Climatic Oscillations

Sudesh Kumar Wadhawan

Abstract Thar Desert has several unique features and shows distinctive assemblage of Quaternary sediment deposits in continental geological basin configurations. A large variety of spectacular dune fields ranging from clustered parabolic to transverse, linear, reticulate, star and barchanoids dunes have been mapped across the vast span of the Thar Desert. The clustered parabolic dunes constitute characteristic aeolian landforms of the Thar Desert that are presently getting disturbed by the contemporary aeolian dynamism and increasing anthropogenic activities. These older stabilized dune forms have been worked upon episodically by aggressive winds and climatic amelioration with increased rainfall that resulted in producing several evolved and superimposed complex dune forms, impregnated with pedogenic calcrete formation. A unique succession of colluvium-aeolian deposits and characteristic stable parabolic and obstacle dunes landscape with other successive and distinctive dune diversity located in the central parts of the Thar Desert are elucidated here as Geoheritage sites of 'sedimentology' and 'geomorphology and active geological processes' in dryland environment that deserve to be preserved for its geoscientific, educational and recreational values and hold immense geotourism potential.

Keywords Thar Desert · Sand dunes · Dune morphology · Colluvio-Aeolian deposits

S. K. Wadhawan (✉)
Geological Survey of India, Center for Wireless Networks and Applications, Amrita Vishwa Vidyapeetham, Amritapuri, Kollam, India

S. C. Tripathi et al. (eds.), *Geoconservation and Geotourism Potential of India*, Society of Earth Scientists Series, https://doi.org/10.1007/978-3-031-81017-6_13

1 Introduction

Indian subcontinent is bestowed with spectacular geodiversity and a variety of geoheritage sites (GSI 2001; Kale 2010; Singh and Anand 2013; Govt. of India 2016; INTACH 2016; Ranawat and Soni 2019; Wadhawan 2020, 2021). A Geosite or unique geologic feature must emphasise its scientific significance into primary consideration in order to be included in the category of geologic heritage or Geoheritage (Brilha 2017). Geotourism is a knowledge-based tourism, an interdisciplinary integration of the tourism industry with conservation and interpretation of abiotic nature attributes. Geotourism adds to ecotourism's principal focus on biodiversity: the plants (flora) and animals (fauna) by adding a third dimension of the abiotic environment—the geodiversity. Thus, it is growing around the world through the growth of UNESCO's Geopark programmes as well as independently in many natural and urban areas where tourism's focus in on the geological environment, landscape and geological processes (UNESCO 1999; Eder 2008; Dowling and Newsome 2010; Dowling 2011; Lazzari and Aloia 2014). Geotourism has proven to be an essential support system for overall regional sustainable development that can leverage the creation of employment and service industry, boosting economic growth and rise in living standards of local communities.

Dryland environments—defined as landscapes spanning hyper-arid to sub-humid climates, cover 40% of the terrestrial surface of earth, host about 40% of the world's human population, and are characterized by considerable seasonal and inter-annual variations in precipitation and typical diurnal temperature fluctuations. Almost one-quarter of the earth's land surface—some 33.7 million square kilometres—has been defined as "desert" in some sense. These deserts are inhabited by over 500 million people, significantly more than previously thought (Dhir et al. 2018). Two major aridity affected tracts occur in Indian subcontinent. These are: the Great Indian sand desert or the Thar Desert—the subtropical hot sandy region in western parts of India and the cold arid region in Ladakh and adjoining Lahul-Spiti in the trans-Himalayan mountains with characteristically low annual precipitation and high evaporation (Fig. 1). The climatic oscillations of the Quaternary period have influenced evolution of the dryland landscapes, ecology, and hominin populations in the Thar Desert. Drivers include glacial to interglacial climatic impacts on temperature and hydrometeorology, processional-insolation forcings of global monsoon systems, and teleconnection with millennial-scale extreme climate events at the higher latitudes. Dryland responses to climatic variation are recorded in landform morphology (geoproxies) and in sedimentary records. The interpretation of aeolian—lacustrine, fluvial, and other dryland archives and sequential records reflect complex and nonlinear arid landscape dynamics and feedbacks in response to dynamic earth system science phenomenon. Locating and studying such rare natural features that preserve the Quaternary climatic oscillations is rewarding exercise for its intrinsic scientific, educational and geotourism values. Present contribution attempts to describe and draw attention to such unique features of rare geological significance the colluvium-aeolian deposits and characteristic stable parabolic dunes landscape with immense

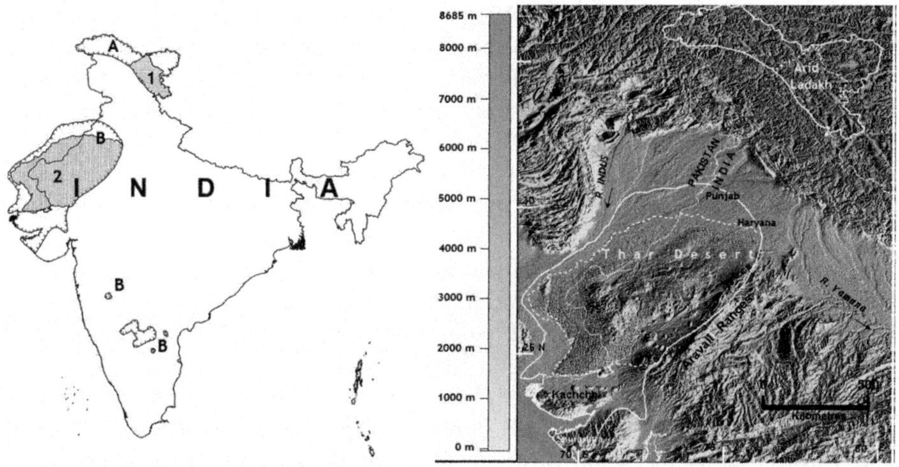

Fig. 1 Extent of Arid Zone in India as viewed in IRS satellite image. Location of hot and cold arid regions. A: Cold arid region; B: Hot arid regions. 1: Ladakh Desert, India; 2: Thar Desert, India and Pakistan. DEM generated image on the right is based on SRTM30-GTOPO data at ~1 km resolution, with annotated boundaries [Image Curtsey Dr. Amal Kar, personal communication]

geotourism potential located in the central parts of the Thar Desert—the dryland environment of the northwest India.

2 Geological and Geomorphological Set-up

The Thar Desert is the world's seventh largest desert on the subtropics. Although it is typically one of the most inhospitable hot and dry ecoregions in the Indo-Pacific, yet it can boast of being the most populated desert. However, 4500–5500 years ago this area supported what is considered to be one of the world's oldest civilizations, the Indus-Saraswati or Mohenjo-Daro and Harappan—Rakhi Giri settlements and related geoarchaeological sites (Valdiya 2002; Michel Danino 2010; Bakshi 2019). The Indian part of the hot and arid Thar Desert is spread over 19.6 Mha in Rajasthan; 6.6 Mha in Gujarat; 1.5 Mha in Punjab and 1.3 Mha in Haryana (Dhir et al. 1992, 2018). Quaternary continental deposits and landforms in Thar Desert have developed over a relatively stable basement platform. Quaternary deposits and landform assemblages provide spectacular landscapes in the Thar Desert that include a diverse variety of dune forms and playa—saline lakes, deflated gravel spreads and wind abraded residual hills of a variety of brightly coloured rock types. Quaternary geological studies affirm that there were discrete and dominant strong aeolian activities in the Thar Desert for the last 200–300 ka period. These were marked by several episodes of greater aridity, stronger wind regime and sand movement followed by periods of stability, implying climatic amelioration, consolidation and formation of calcareous soils/calcretes. Wetter episodes accumulated colluvial deposits skirting the residual hills and fluvial sand sheets and the coastal sediments that eventually acted as provenance or the source material for southwest monsoon winds to work on during prolonged dry spells and generate a variety of dunes in different sedimentation sub-basins with distinctive substrata in the Thar Desert. Such typical sites as are located in the central part of Thar Desert near the Shergarh Trijunction in Jodhpur district, Rajasthan, assume greater significance and qualify for getting included in the coveted list of Geoheritage sites that would need to be conserved and utilised for geotourism purposes.

Thar Desert is confined essentially to West Rajasthan Shelf (WRS). The tectonic disposition and basement configuration of intra-cratonic basins and sedimentary formations here range from the highly deformed rocks of the Precambrian Aravalli Fold Belt (3.5–2.5 Ga) and the Delhi Supergroups (~1 Ga), as part of the Great Indian Proterozoic Fold Belts along the western edge of the ancient Rodinia Supercontinent (Ashwal et al. 2013). A series of igneous events between 1700 Ma and 750 Ma along the south-western margin of the Aravalli Hills produced the Sendra Granite, Erinpura Granite, Malani Igneous Suite, Mt. Abu Granitoids, etc., in central Rajasthan (Meert et al. 2010); to undeformed sedimentary rocks of Late Proterozoic to Early Palaeozoic—the Marwar Supergroup of sedimentary rocks in the middle, to Mesozoic and Cainozoic cover sedimentary formations on the western and north-western fringe areas in western parts of Rajasthan and Haryana in NW India. Based on detailed field surveys, geological mapping of the Quaternary formations, their nature of surface disposition and inferred sub-surface configuration in arid-semiarid western parts of India, it has been feasible to delineate a few large and broadly linear Neogene sub-basins or tectonic depressions filled-in with aggradational continental Quaternary sediments that appear separated respectively by the erosional relief features. A series of north-easterly trending linear stepped graben structures have been deciphered showing westerly downthrown blocks respectively (Fig. 2; Wadhawan 2010, 2018; Dhir et al. 1992).

It is remarkable that the geospatial distributions of these pre-Quaternary rocks show a successive geologic younger age from the eastern margin of the Thar along the Aravalli Hills, to the west in Jaisalmer region. A schematic profile, say along 27°N also depicts a general fall in average elevation from 500 m—accordant summit levels of the Aravalli hill range on the east to 80–120 m above msl on the western part of the Thar Desert that forms the prevailing core of the active sands covered desert (Fig. 2). This appears to be in direct correspondence to the prevailing climatic conditions of extreme aridity on the west that grades gradually and becomes semiarid with increasing rainfall towards the east.

Quaternary sedimentation history in each continental sub-basin is unique, but broadly comparable and multicyclic and varying in degree of development: starting with coarse colluvium/ill-sorted polymictic fluvial clastic deposits, through silt-clays of playa/lacustrine deposits and culminating with sandy aeolian deposits. It is established that the fluvial (river born) and fluvio-lacustrine deposits form the base for aeolian deposition and constrain the overall configuration and disposition of older aeolian deposits (dune fields) in the Thar Desert (Wadhawan 2018, 2023). Large scale drainage disruption/segmentation in the Thar Desert is inferred either

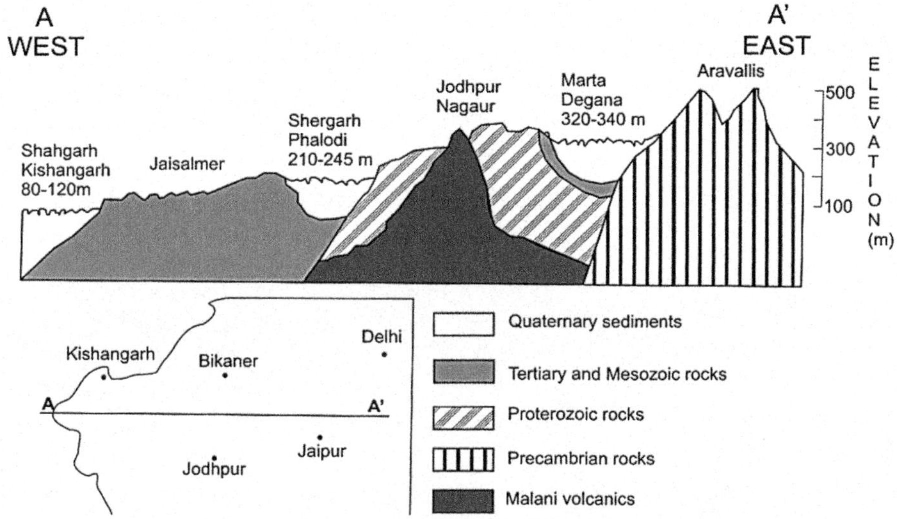

Fig. 2 Schematic profile along 27°North showing Neogene/Quaternary Sub-basins in the Thar Desert (After Wadhawan 2010)

due to neotectonics or obstruction of rivers channels by extending dune fields driven by the southwest winds, for instance present day desiccated Ghaggar River along the northern periphery of the Thar Desert.

A large variety of distinctive dune fields ranging from clustered parabolic to transverse, linear, reticulate, the star and barchanoids have been mapped across the vast span of the Thar Desert (Fig. 3). The characteristic aeolian landforms of the Thar Desert are the stabilised clustered parabolic dunes that constitute the present-day active desert front experiencing reactivation of the sandy crestal parts owing to contemporary aeolian dynamism. These older stabilized dune forms have been worked upon by winds episodically that resulted in producing several evolved and superimposed complex dune forms (Fig. 4). Morpho-stratigraphically, episodic natures of multi-phase aeolian deposits are distinguished in the central Thar Desert that rest over shallow fluvial and calcretised colluvial sedimentary units. However, the buried Neogene fluvio-lacustrine continental deposits in the northern and north-western parts of the Thar Desert show exceptionally large thickness (Saini et al. 2009; Saini and Mujtaba 2012; Wadhawan 2018).

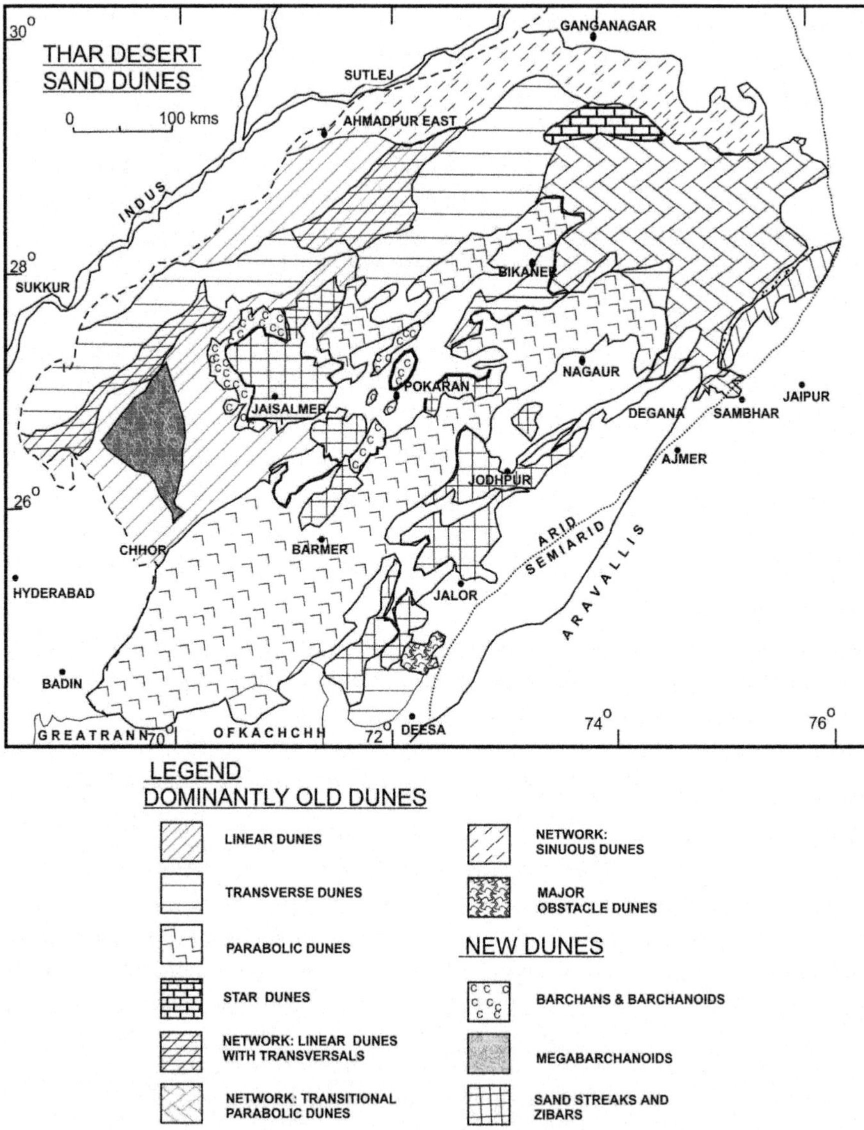

Fig. 3 Distribution and morphology of the aeolian sand deposits/Dunes in the Great Thar Desert (After Singhvi and Kar 2004)

3 Geoheritage Site of Colluvio-Aeolian Sequence Near Shergarh Trijunction

This site near Shergarh Trijunction is located about 70 km west of Jodhpur also called locally as Khirja Fatehpur, falling in Survey of India Toposheet No. 45B/6

Fig. 4 A bird's eye view of the residual hills of the purple rhyolite rock of the Malani Igneous Suit and colluvium/slope-wash of angular gravelly spread at places buried under older obstacle dunes deposits and recently reactivated golden yellow sand aggradations around Shergarh Trijunction

(1:50,000 scale). The Colluvio-Aeolian sequence occurs on a pediment foot-slope of a prominent hill comprising Precambrian rhyolites of Malani Igneous Suite, occurring dominantly in west-central Rajasthan, with an elevation of about 55 m above the ambient plains. The residual hills of the purple rhyolite rock of the Makani Igneous Suit and colluvium/slope wash with older stabilized obstacle dunes—parabolic dunes and recently reactivated sand aggradations provide a spectacular landscape in desert environment (Fig. 4).

The present section is exposed as a result of quarrying for calcrete and gravels. The section is remarkable in the sense that despite the small size of the pediment, the aeolian and colluvial deposits skirting the hill are distinctly separable from each other and occur in horizontally disposed alternating beds. In the four-meter-thick sequence, there are four distinct colluvial deposit members, made up of fresh-looking angular gravels and coarse angular fragments (2–5 cm) of red-purple rhyolite, at places coated and cemented by lime- calcium carbonate (Figs. 5, 6 and 7). The intervening aeolian sand bed is typically similar to the surrounding dune field. The sands are fine grained well-sorted and calcareous in nature. Caliche is a reddish-brown to white layer found in many desert soils. Caliche or clayey *Kankar* deposits commonly occurs as nodules or as coatings on mineral grains formed by the complicated interaction between water and carbon dioxide released by plant roots or by decaying organic material. The lime here has been redistributed into 1–3 cm size nodules (*Kankars*) comprising fine aggregated sand and micrite calcite. The gravels are coated with calcium carbonate and locally cemented. Occasional occurrences of rhizolith nodules are observed. The coarse sized colluvial clastic sediments amidst aeolian sands are a clear indication of changes in the environment of deposition. Absence of any appreciable aeolian contamination in the colluvial members clearly suggests that the adjoining landforms were stable during the period of the deposition of colluvial members. Later, sand sheets I to IV were dated using luminescence method (Infrared Stimulated Luminescence on feldspars and Blue Green Stimulated

a b

Fig. 5 (**a** and **b**) Colluvio-Aeolian deposits in Thar Desert exposed near Shergarh Trijunction in quarries for road building material. Note the gentle dip-slope on the left (**a**), at proximal end near the residual hill and flattening to nearly horizontal disposition at the distal end on the right (**b**)

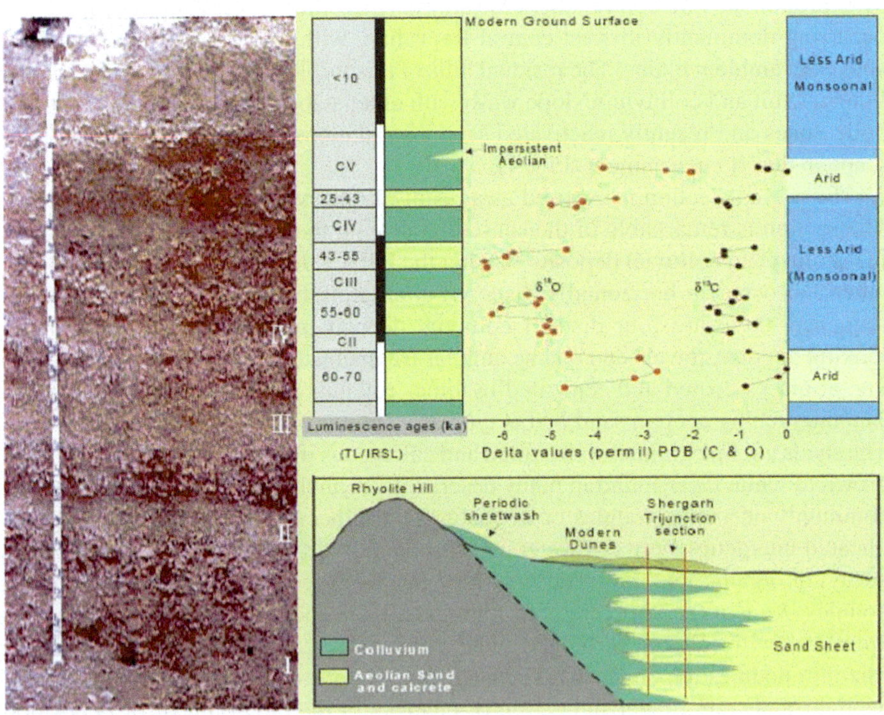

Fig. 6 Exposed Colluvium-aeolian section with absolute dates and isotopic details and paleoclimatic interpretations; Oxygen and Carbon Isotopic data is after Andrews et al. 1998

Fig. 7 Stabilised clustered parabolic dunes in central parts of the Thar Desert showing reactivation of crestal parts and recent sand mobilisation

Luminescence on Quartz) that provided absolute dates of 60–70, 55–60, 45–55 and 25–45 ka respectively (Andrews et al. 1998). Analysis of calcretes (Figs. 5a, b), shows that its δ13C values range from 0 to - 1.8 0/00 whereas the δ18O values showed a larger range with values from −2.1 to −6.5 0/00.

3.1 Description of the Exposed Colluvium-Aeolian Section

The long section exposed in the excavated shallow quarry showed four sets of colluviums and aeolian sand deposits that were logged and details on the sedimentation recorded. The deposits showed conformity to a gentle dip-slope away from the residual hill in close proximity becoming nearly horizontal about 200 m away. These are labelled as Colluvium-1 to IV and Aeolian-I to V, respectively from bottom of the excavated section to the top (Fig. 6).

3.2 Geoheritage Site of Colluvium-Aeolian Section Details

0–110 cm, AEOLIAN V: Light yellowish brown (10 YR6/4), fine sandy, structureless, grading to a weak, subangular blocky, friable, semi consolidated structure. The top 40 cm is weakly calcareous with up to 1%$CaCO_3$ content increases with depth to about 5% at the bottom of the members. Sharp transition to.

110–160 cm, COLLUVIUM V: Greyish white (10 YR 7.5/2), locally pale brown (10YR 7/3). Principally made up of 1–10 cm angular fragments of rhyolite, coated with lime. Some aeolian lenses are also present. Sharp transition to,

160–180 cm, AEOLIAN IV: Pale brown (10 YR 6/3), loamy fine sand with weak subangular blocky structure, having weakly formed and friable lime nodules (~15–20% CaCO–) that are randomly distributed in the matrix. The nodules, upon fracturing, show an amorphous appearance, though the core in some cases is crystalline. Sharp transition to,

180–220 cm. COLLUVIUM IV: Greyish white (10 YR 7.5/2), angular rhyolite fragments, mostly 1–10 cm, poorly sorted, thinly coated with lime. Some fine nodules of lime are present. Sharp transition to,

220–240 cm. AEOLIAN III: Similar to Aeolian II, except that lime nodules are more abundant (20–30%). They are hard and upon breaking show a moderately crystalline fabric.

240–260 cm. COLLUVIUM III: Similar to Colluvium IV.

260–290 cm. AEOLIAN II: Similar to Aeolian III, except that the lime nodules are bigger and also more frequent. Upon fracturing, the nodules show an amorphous to partly crystalline calcite fabric, amidst detrital matter. The detrital matter in the nodules constitutes 36% of the total, while Ca as $CaCO_3$ constitutes 62%, with traces of magnesium, sodium and potassium.

290–315 cm. COLLUVIUM II: Similar to Colluvium III.

315–370 cm. AEOLIAN I: Similar to Aeolian III, except that the lime nodules are few.

>370 cm. COLLUVIUM I: Coarse angular clastics of rhyolites bedrock and in situ bed rock of the Malani Igneous Suite of Volcanic rocks (at the base, not fully exposed in the section).

It is recorded that calcrete nodules occurs in the aforesaid four laterally persistent and one nonpersistent aeolian units, separated by colluvial gravel. Andrews et al. (1998) analysed oxygen and carbon stable isotopes in the pedogenic calcrete nodules developed in the aeolian units of this Quaternary sediments of colluvio-aeolian sequence of the Thar Desert in Rajasthan. Luminescence dating provides a chronometric perspective to aeolian dynamics and dune sand accumulation phase, which has been used to reconstruct the past geomorphological changes or the dune building successions in the Thar Desert against a backdrop of strengthening and weakening of Indian Summer Monsoon (Singhvi and Kar 2004). Andrews et al. (1998) also provided the chronological dates for the deposition of these Aeolian units I to V, as encountered at the Shergarh Trijunction. Thermoluminescence and infrared- and green-light-stimulated luminescence of host quartz and feldspar grains gave age brackets for persistent Aeolian units I to IV of ca. 70,000–60,000, ca. 60,000–55,000, ca. 55,000–43,000, and ca. 43,000–~25,000 years, respectively. However, the topmost youngest Aeolian unit V, is <10,000 years old and contains no calcrete.

Here the Late Quaternary climatic oscillations were interpreted from the analytical data of stable oxygen isotope compositions of calcretes in terms of intensity of monsoonal rainfalls as: in most of Aeolian unit I, in the upper part of Aeolian unit IV, and in the nonpersistent aeolian unit, range between −4.6 and −2.1‰ PDB. These values, up to 4.4‰ greater than values from Aeolian units II and III, were interpreted as representing non-monsoonal ^{18}O-enriched "normal continental" waters during climatic phases when the monsoon weakened or failed. On the other hand, 25,000–60,000-year-old calcretes (Aeolian units II and III) probably formed under monsoonal conditions. The two periods of weakened monsoon are consistent with other paleoclimatic data from India and may represent widespread aridity on the Indian subcontinent during Isotope Stages 2 and 4 (Dhir et al. 2018). The total variation in $\delta^{13}C$ is 1.7‰ (0.0–1.7‰), and $\delta^{13}C$ covaries positively and linearly with $\delta^{18}O$. $\delta^{13}C$ values are highest when $\delta^{18}O$ values indicate the most arid climatic conditions. This is best explained by expansion of C_4 grasses at the expense of C_3 plants at low latitudes during glacial periods when atmospheric pCO_2 was lowered. C_4 dominance was majorly influenced by global change in atmospheric pCO_2 despite the lowered summer rainfall.

4 Hill Top Panoramic View of Stable and Reactivated Parabolic Dunes Landscape

The residual hill top near the Quaternary Colluvio-Aeolian deposits at Shergarh Trijunction provides a typical panoramic view of the series of the characteristic clustered parabolic dunes and obstacle dunes in the central part of the Thar Desert (Figs. 4 and 7). A walking track or cemented ramp goes up to the hill top where a Shiva Temple is also built and maintained that is visited frequently by the devotees. U-shaped mounds of aeolian sand with convex noses trailed by elongated arms are parabolic dunes. These dunes are formed from blowout dunes where the erosion of vegetated sand leads to a U-shaped depression. The elongated arms are held in place by vegetation; the largest arm is recorded in Thar Desert to be up to 8 km. The dominant northeasterly orientation of the long arms of the parabolic dunes indicates their formation by the dominant SW wind. Parabolic dunes occur in areas where very strong winds are mostly unidirectional and fairly consistent as is the case with the parabolic dune fields in central and south-western Thar Desert. Although these dunes were found in areas now characterized by moderately variable seasonally high wind speeds, the effective winds associated with the growth and migration of both the parabolic and crescent dunes are observed to be the most consistent in south-westerly wind direction.

The dune height varies from 15 to 50 m. Each dune is linked to an adjacent one, so that they appear to share arms, although the individual dunes override one another in a manner analogous to the generally observed coastal parabolics (Wadhawan 2023). The dune arms begin from the sandy plain and steadily gain height downwind along with the sandy corridors in between. These converge in a sharp curve. Near to the nosal part of the parabolic dune, the sandy corridors become gradually constricted and show a common deflationary blow-out feature of the dune. The windward slope is about 5°–10°and the slip face slope varies from 24° to 30°. Both the arms of the parabolic dune have a steeper (25°–30°) outward slope and a gentler (10°–15°) inward slope. The grain size for these well-sorted, very fine to medium sands is about 0.06–0.5 mm. The crestal region of the dune in the area is dominated by the well-sorted fine sand grain size 0.15 mm, although at a depth of about 150 cm, the dominant sand grain size is 0.13 mm (Wadhawan 1996). Presently reactivated crestal parts of the older stable parabolic dunes support superimposed aeolian sands in the form of linear dunes as obstacle extensions or climbing and falling barchans thus forming a complex multi-storied aeolian deposits (Fig. 7). Synthesis of dune trends in all across the Thar desert, particularly the orientation of the longitudinal elements of the dunes suggests that the overall sand moving winds remained moving from SW to NE implying that the causative monsoon (ISM) atmospheric circulation patterns have remained persistent and fairly stable since at least the Late Quaternary period. These Quaternary continental basins support distinctive aeolian landforms that have evolved as per the prevailing climate gradient of decreasing aridity from west to east and corresponding increase in rainfall. The occurrence of characteristic stabilized parabolic dunes of clustered or rake-like disposition and derivative linear dunes comprising fine well-sorted sand and consolidated with pedo-calcification show superimposed multi-tiered arrangement of complex dunes with higher relief thereby representing different episodes of aeolian dynamism and dune morphologic evolution.

Preliminary TL dates on the top 3 m of sand from three locations on a dune suggest two periods of accretion, around 12 ka and 4 ka respectively. The sand dunes and interdune plains support a variety of plant species like *Prosopis cineraria, Prosopis juliflora, Acacia jacquemontii, Calotropis procera, Aerva tomentosa and Crotalaria burhia.* However, increased human activities around the settlements, deep ploughing along the dune slopes and interdune plains and destruction of vegetarian cover are leading to reactivation of sand with consequent formation of new, mobile bedforms (Figs. 4 and 7). As a consequence, the dune crests and flanks often have 1–2 m high barchanoids and loose sand accumulation. Considering the adverse effects of such sand mobility on motor ways and settlements, a sand stabilization programme in the region has been undertaken by the State through plantation of *Acacia tortilis* (Dhir et al. 2018).

Dhir et al. (2018) showed the oldest aeolian activity at 200–250 ka in the basal part of an aeolian sequence at Chamu, in central Thar desert. Besides, several younger episodes dated to about 150, 90–100, about 60, 25–30 and subsequent vigorous aeolian dynamism was during 17–11 ka with most widespread dated event at about 13 ka, when monsoon winds were re-establishing and not at the peak of aridity during Last Glacial Maxima (Singhvi and Kar 2004). These aeolian events were separated by landscape stability, soils and calcretes formation indicating climate amelioration. Calcretes are secondary, mostly calcium carbonate accumulations whose formation involves dissolution, transport and accumulation at some depth in soil profile of the carbonate already present in a thickness of sediments or that brought in solution by runoff water or as sediment/dust from surrounding landscape. These are typical formations of a semi-arid climate, where rainfall during a part of the year is sufficient enough to cause mobilization of carbonate as percolating soil solutions to be followed by a period of dryness for dissolved material to precipitate at a depth. The calcretes have stages of development or maturity. The initial stage is presence of carbonate as loose, porous segregations with diffuse boundary with the surrounding matrix. More advanced stage is formation of chalky and then of hard and complex nodules with, a distinct boundary. With further carbonate enrichment, the nodules coalesce to form spatially dense, compact and hard strata with continuous carbonate phase and this is often called a hardpan.

5 Thar Desert Landscape of Dunes Diversity: Stabilised Dunes with Superimposed Active Barchanoids Around Sam–Dhanana in Jaisalmer Basin

Sam–Dhanana area is located about 50 km west-southwest of golden-yellow city of Jaisalmer, in a region with ~150 mm annual precipitation. This is a famous tourist spot for its aesthetic view of active and clean dunes sands, camel safaris and sunset-view points in the western part of central Thar Desert (Fig. 8, Google Earth Image). The dunes landscape here displays hair-pin shaped stabilised parabolic dunes and complex longitudinal dunes transitional to the parabolic dunes. The height of these dunes ranges from 10 to 25 m and length up to 10 km. Although the dunes are stable with extensive calcareous nodules and fossil root casts, a part of them is barren and has chains of active barchanoids dunes with their scenic geometric patterns (Fig. 9: Field Photos a-e). However, this is also the most mobile part of the dune fields, engulfing newer territories each year as the summer hot blowing wind is seasonally very strong. The sands are well sorted, fine skewed and feebly calcareous. The mean grain size is 0.16 mm in the case of stable surface and 0.19 mm for the reactivated barchanoids dunes at the top. The isolated barchans, encountered to the west of the village, have still coarse mean grain size and the sorting is moderate. As most of the territory falls under the protected Desert National Park, the dune stabilisation programs over the reactivated dunes in this region are being implemented and monitored by government agencies. Mulching was done with Sewan Grass and other bushes—vegetation that thrives in extreme dryland environment. Grasses and bush plants such as *Lasiurus sindicus and Crotalaria burbi, Acacia tortilis* were grown in rows within the mulches. Reseeding was done with *Lasiurus sindicus. Haloxylon*

Fig. 8 Satellite image showing Tourist Camping sites and series of superimposed active barchanoids dunes over stabilised older linear dunes at Sam. Southwestern part of the dune field shows efforts for dunes stabilisation by plantations of dryland vegetation and grasses

salicornicum, Aerva persica and *Leptadenia pyrotechnica* are the other dryland plant species that were planted (Fig. 8).

Human implements were found here in the form of micro-blade cores, retouched blades and debitage on the stabilized surface, implying thereby that these dunes predate the Mesolithic period (~10 ka–15 ka BP). A preliminary TL date of about 15 ka BP on a sample from a depth of 0.8 m in stabilized dune, also confirms this archaeological evidence. Recent TL dating of series of fossil dunes in the Thar Desert suggests that the dune building activity was at its peak around 14 ka BP when the SW Indian Summer Monsoon was gradually gaining momentum. Archaeological data is also in conformity with this inference (Dhir et al. 1992, 2010, 2018).

6 Concluding Remarks

Conservation of rare geological entities that are uniquely significant from geological, geomorphological values are important from geoscience education and recreation points of view. These are referred to as geoheritage sites that also have intrinsic potential for development of geotourism. Thus, geotourism is the promotion of tourism at geosites along with the conservation of geodiversity with an understanding of earth sciences through scientific interpretation and learning as achieved through physical visits to the identified geological features.

Quaternary Colluvio-Aeolian deposits and unique Parabolic Dunes Landscape in Central Thar Desert of western India are elucidated here to emphasise their

Fig. 9 (**a**) Satellite imagery showing complex stabilised longitudinal dune field superimposed with younger and active barchans, 40 km WSW of Jaisalmer; (**b**) satellite imagery showing hair-pin shaped stabilised parabolic dunes in Jaisalmer basin, Thar Desert; (**c**) and (**d**) field photos show active sand dunes at Sam in the western parts of Jaisalmer Basin; e mobile sand dunes and golden aeolian sands viewed from atop the active dunes at Dhanana in the Jaisalmer Basin, Thar Desert (Field Photos after Wadhawan 1994, 2023)

geotourism potential and implied need for geoconservation. Presently recorded studies of such rare natural features of colluvium-aeolian deposits that preserve the Quaternary climatic oscillations is rewarding exercise for its immense intrinsic geo-scientific, educational and geotourism values.

The diversity of dunes landscape showing spectacular geometry of golden-yellow clean sand deposits of different generations in the central Thar Desert are the distinctive value-added tourist attractions for recreation and geoscientific education purposes. These geoheritage sites represent 'Sedimentology' and 'Geomorphology and active geological processes' in dryland environment. These geoheritage sites are easily accessible and frequently visited by national and international delegates, researchers and geoscience enthusiasts. Therefore, it imperative to adopt a systematic and strategic approach towards the protection of geoheritage sites and promotion of enhanced and sustainable development of the geotourism industry.

With the creation of basic civic amenities and explanatory display boards, these rare geoheritage sites have great scientific value as outdoor classrooms for enhancing public understanding of the natural geological phenomena, processes and Late Quaternary climatic oscillations preserved in sedimentary records showing post-depositional modification (formation of calcrete within porous aeolian sand deposits) under landscape stability and increased rainfall conditions in otherwise dryland environment of Thar Desert.

References

Andrews JE, Singhvi AK, Kailath AJ, Kuhn R, Dennis PF, Tandon SK, Dhir RP (1998) Do stable isotope data from Calcrete record late Pleistocene monsoonal climate variation in the Thar Desert of India? Quat Res 50(3):240–251. https://doi.org/10.1006/qres.1998.2002

Ashwal LD, Solanki AM, Pandit MK, Corfu F, Hendriks BWH, Burke K, Torsvik TH (2013) Geochronology and geochemistry of Neoproterozoic Mt. Abu granitoids, NW India: regional correlation and implications for Rodina paleogeography. Precambrian Res 236:265–281

Bakshi GD (2019) The Sarasvati civilisation—a paradigm shift in ancient Indian history. Garuda Publications, Bharat. 268 p

Brilha JB (2017) Geodiversity and geoconservation. Geoheritage 12. https://doi.org/10.1007/s12371-020-00475-2

Brocx M, Semeniuk V (2007) Geoheritage and geoconservation - history, definition, scope and scale. J R Soc West Austr 90:53–87

Dhir RP, Kar A, Wadhawan SK, Rajaguru SN, Misra VN, Singhvi AK, Sharma SB (1992) Thar Desert in Rajasthan - land, man and environment. Geological Society of India, Bangalore, p 191p

Dhir RP, Singhvi AK, Andrews JE, Kar A, Sareen BK, Tandon SK, Kailath A, Thomas JV (2010) Multiple episodes of aggradation and calcrete formation in late Quaternary aeolian sands, Central Thar Desert, Rajasthan, India. J Asian Earth Sci 37:10–16

Dhir RP, Joshi DC, Kathju S (2018) Thar desert - in retrospect and prospect. Scientific Publishers, Jodhpur, p 405p

Dowling RK (2011) Geotourism's global growth. Geoheritage 3(1):1–13

Dowling RK, Newsome D (2010) Global geotourism perspectives. Goodfellow Publishers, Oxford

Eder W (2008) Geoparks—promotion of earth sciences through Geoheritage conservation, education and tourism. J Geol Soc India, Bangaluru 72:149–154

Geological Survey of India (2001) National Geological Monuments (Anantharamu, T.R., Bellur, D., Bhasker, A.A., Eds.). Geological Survey of India, Special Publication No. 61, 98 p

Government of India (2016) Geotourism hotspots of Indian subcontinent, 36th international geological congress (IGC). Geological Survey of India, New Delhi, p 101p

Henriques MH, Pena dos Reis R, Brilha J, Mota TS (2011) Geoconservation as an emerging geoscience. Geoheritage 3(2):117–128

Hose TA (2012) 3G's for modern Geotourism. Geoheritage 4:7–24

Indian National Trust for Art and Cultural Heritage [INTACH] (2016) A monograph on National Geoheritage Monuments of India. Natural Heritage Division, New Delhi. 146 p

Kale VS (2010) The Western Ghats: the great escarpment of India. In: Migon P (ed) Geomorphological landscapes of the world. Springer, New York, pp 257–264

Lazzari M, Aloia A (2014) Geoparks, geoheritage and geotourism: opportunities and tools in sustainable development of the territory. Geo J Tour Geosites [GTG] 13(1):110–123

Meert JG, Pandit MK, Pradhan VR, Banks J, Sirianni R, Stoud M, Newstead B, Gifford J (2010) Precambrian crustal evolution of peninsular India: a 3.0-billion-year odyssey. J Asian Earth Sci 39:483–515

Michel Danino (2010) The lost river- on the trail of the Sarasvati. Penguin Random House. 357 p

Ranawat PS, Soni G (2019) Potential geoheritage & geotourism sites in India. Int J Sci Res Pub 9(6):91–96

Saini HS, Mujtaba SAI (2012) Depositional history and paleoclimatic variations at the northeastern fringe of Thar Desert, Haryana plains, India. Quat Int 250:37–48. https://doi.org/10.1016/j.quaint.2011.06.002

Saini HS, Tandon SK, Mujtaba SAI, Pant NC, Khorana RK (2009) Reconstruction of buried-floodplain systems of the NW Haryana Plains and their relation to the 'Vedic' Sarasvati. Curr Sci 97:1634–1643
Singh RB, Anand S (2013) Geodiversity, geographical heritage and geoparks in India. Intl Jour Geoheritage 1:10–26
Singhvi AK, Kar A (2004) The aeolian sedimentation record of the Thar Desert. Proc Indian Acad Sci (Earth Planetary Sci) 113:371–401
UNESCO (1999) UNESCO Geoparks Programme—a new initiative to promote a global network of geoparks safeguarding and developing selected areas having significant geological features: document 156 EX/11 rev., executive board, 156th session. UNESCO, Paris, 4 p
Valdiya KS (2002) Sarasvati: the river that disappeared. Universities Press, Hyderabad. 116 p
Wadhawan SK (1994) Dune dynamics and evolution of aeolian landforms in parts of Jaisalmer district, Rajasthan. J Indian Soc Remote Sens 22(2):65–77
Wadhawan SK (1996) Textural attributes of recent Aeolian deposits in different sub-basins of the Thar Desert, India. J Arid Environ 32:37–51
Wadhawan SK (2010) Quaternary continental sedimentation basins and dune fields in Thar Desert of NW India. Gondwana Geol Magaz 12:265–270
Wadhawan SK (2018) Palaeogene—Neogene tectonics and continental aggradational basins in North-Western India: implications for geological evolution of Thar Desert. In: Sunil Bajpai SC, Tripathi, Prasad V (eds) The Indian Paleogene. Springer, pp 151–166
Wadhawan SK (2020) Geoheritage sites and potential geoparks in India strategies for sustainable development. Indian J Geosci 74(3):354–362
Wadhawan SK (2021) Geoheritage and potential geotourism in geoparks-Indian perspective. In: Singh R, Wei D, Anand S (eds) Global geographical heritage, geoparks and geotourism. Advances in geographical and environmental sciences. Springer, Singapore. https://doi.org/10.1007/978-981-15-4956-4_15
Wadhawan SK (2023) Geospatial Diversity, Geologic Significance and Evolutionary Trends of Parabolic Dunes in Thar Desert of the Indian Subcontinent. In: Qi L, Gaur MK, Squires VR (eds) Sand Dunes of The Northern Hemisphere - distribution, formation, migration and management, vol 1: Formation and Evolution of Dune Landscapes. Taylor & Francis Group, CRC Press, pp 297–314

The manufacturer's authorised representative in the EU is Springer Nature Customer Service Centre GmbH, Europaplatz 3, 69115 Heidelberg, Germany. If you have any concerns regarding our products, please contact ProductSafety@springernature.com

Printed and bound by CPI Group (UK) Ltd, Croydon, CR0 4YY

30/09/2026

02225571-0002